Pre-Algebra Workbook

Essential Learning Math Skills Plus Two Pre-Algebra Practice Tests

By

Michael Smith & Reza Nazari

Pre-Algebra Workbook

Published in the United State of America By

The Math Notion

Web: WWW.MathNotion.Com

Email: info@Mathnotion.com

Copyright © 2020 by the Math Notion. All rights reserved. No part of this publication may be reproduced, stored in a retrieval system, or transmitted in any form or by any means, electronic, mechanical, photocopying, recording, scanning, or otherwise, except as permitted under Section 107 or 108 of the 1976 United States Copyright Ac, without permission of the author.

All inquiries should be addressed to the Math Notion.

About the Author

Michael Smith has been a math instructor for over a decade now. He holds a master's degree in Management. Since 2006, Michael has devoted his time to both teaching and developing exceptional math learning materials. As a Math instructor and test prep expert, Michael has worked with thousands of students. He has used the feedback of his students to develop a unique study program that can be used by students to drastically improve their math score fast and effectively.

- **– SAT Math Practice Book**
- **– ACT Math Practice Book**
- **– PSAT Math Practice Book**
- **– Algebra Math Practice Books**
- **– Common Core Math Practice Books**
- **–many Math Education Workbooks, Exercise Books and Study Guides**

As an experienced Math teacher, Mr. Smith employs a variety of formats to help students achieve their goals: He tutors online and in person, he teaches students in large groups, and he provides training materials and textbooks through his website and through Amazon.

You can contact Michael via email at:
info@Mathnotion.com

Prepare for the Pre-Algebra with a Perfect Workbook!

Pre-Algebra Workbook is a learning workbook to prevent learning loss. It helps you retain and strengthen your Math skills and provides a strong foundation for success. This Pre-Algebra book provides you with solid foundation to get a head starts on your upcoming Pre-Algebra Test.

Pre-Algebra Workbook is designed by top math instructors to help students prepare for the Pre-Algebra course. It provides students with an in-depth focus on the Pre-Algebra concepts. This is a prestigious resource for those who need an extra practice to succeed on the Pre-Algebra test.

Pre-Algebra Workbook contains many exciting and unique features to help you score higher on the Pre-Algebra test, including:

- Over 2,500 Pre-Algebra Practice questions with answers
- Complete coverage of all Math concepts which students will need to ace the Pre-Algebra test
- Two Pre-Algebra practice tests with detailed answers
- Content 100% aligned with the latest Pre-Algebra courses

This Comprehensive Workbook for the Pre-Algebra is a perfect resource for those Pre-Algebra takers who want to review core content areas, brush-up in math, discover their strengths and weaknesses, and achieve their best scores on the Pre-Algebra test.

WWW.MathNotion.COM

… So Much More Online!

- ✓ FREE Math Lessons

- ✓ More Math Learning Books!

- ✓ Mathematics Worksheets

- ✓ Online Math Tutors

For a PDF Version of This Book

Please Visit WWW.MathNotion.com

contents

Chapter 1: Integers and Number Theory 11

    Whole Number Addition and Subtraction 12

    Whole Number Multiplication and Division 13

    Rounding and Estimates 14

    Adding and Subtracting Integers 15

    Multiplying and Dividing Integers 16

    Order of Operations 17

    Ordering Integers and Numbers 18

    Integers and Absolute Value 19

    Factoring Numbers 20

    Greatest Common Factor 21

    Least Common Multiple 22

    Answers of Worksheets – Chapter 1 23

Chapter 2: Fractions and Decimals 26

    Simplifying Fractions 27

    Adding and Subtracting Fractions 28

    Multiplying and Dividing Fractions 29

    Adding and Subtracting Mixed Numbers 30

    Multiplying and Dividing Mixed Numbers 31

    Adding and Subtracting Decimals 32

    Multiplying and Dividing Decimals 33

    Comparing Decimals 34

    Rounding Decimals 35

    Answers of Worksheets – Chapter 2 36

Chapter 3: Proportions, Ratios, and Percent 39

    Simplifying Ratios 40

    Proportional Ratios 41

    Similarity and Ratios 42

Ratio and Rates Word Problems 43

Percentage Calculations 44

Percent Problems 45

Discount, Tax and Tip 46

Percent of Change 47

Simple Interest 48

Answers of Worksheets – Chapter 3 49

Chapter 4: Exponents and Radicals Expressions 53

Multiplication Property of Exponents 54

Zero and Negative Exponents 55

Division Property of Exponents 56

Powers of Products and Quotients 57

Negative Exponents and Negative Bases 58

Scientific Notation 59

Square Roots 60

Simplifying Radical Expressions 61

Answers of Worksheets – Chapter 4 62

Chapter 5: Algebraic Expressions 65

Simplifying Variable Expressions 66

Simplifying Polynomial Expressions 67

Translate Phrases into an Algebraic Statement 68

The Distributive Property 69

Evaluating One Variable Expressions 70

Evaluating Two Variables Expressions 71

Combining like Terms 72

Answers of Worksheets – Chapter 5 73

Chapter 6: Equations and Inequalities 75

One–Step Equations 76

Multi–Step Equations 77

Graphing Single–Variable Inequalities 78

One–Step Inequalities 79

Multi-Step Inequalities 80

Systems of Equations 81

Systems of Equations Word Problems 82
Answers of Worksheets – Chapter 6 83

Chapter 7: Polynomials 87

Writing Polynomials in Standard Form 88
Simplifying Polynomials 89
Adding and Subtracting Polynomials 90
Multiplying Monomials 91
Multiplying and Dividing Monomials 92
Multiplying a Polynomial and a Monomial 93
Multiplying Binomials 94
Factoring Trinomials 95
Operations with Polynomials 96
Answers of Worksheets – Chapter 7 97

Chapter 8: Geometry and Solid Figures 101

Angles 102
Pythagorean Relationship 103
Triangles 104
Polygons 105
Trapezoids 106
Circles 107
Cubes 108
Rectangular Prism 109
Cylinder 110
Pyramids and Cone 111
Answers of Worksheets – Chapter 8 112

Chapter 9: Statistics and Probability 115

Mean and Median 116
Mode and Range 117
Times Series 118
Stem–and–Leaf Plot 119
Pie Graph 120
Probability Problems 121
Answers of Worksheets – Chapter 9 122

Pre-Algebra Practice Tests... 125
Pre-Algebra Practice Test 1 129
Pre-Algebra Practice Test 2 141
Answers and Explanations... 155
Answer Key...155
Practice Test 1 157
Practice Test 2 165

Chapter 1: Integers and Number Theory

Topics that you will practice in this chapter:

- ✓ Whole Number Addition and Subtraction
- ✓ Whole Number Multiplication and Division
- ✓ Rounding and Estimates
- ✓ Adding and Subtracting Integers
- ✓ Multiplying and Dividing Integers
- ✓ Order of Operations
- ✓ Ordering Integers and Numbers
- ✓ Integers and Absolute Value
- ✓ Factoring Numbers
- ✓ Greatest Common Factor (GCF)
- ✓ Least Common Multiple (LCM)

"Wherever there is number, there is beauty." −Proclus

Whole Number Addition and Subtraction

✎ Find the sum or subtract.

1) $\begin{array}{r} 1,982 \\ +895 \\ \hline \rule{1cm}{0.4pt} \end{array}$

5) $\begin{array}{r} 1,125 \\ +859.35 \\ \hline \rule{1cm}{0.4pt} \end{array}$

9) $\begin{array}{r} 7,322 \\ -895.9 \\ \hline \rule{1cm}{0.4pt} \end{array}$

2) $\begin{array}{r} 3,658 \\ -1,254 \\ \hline \rule{1cm}{0.4pt} \end{array}$

6) $\begin{array}{r} 857.26 \\ +989.15 \\ \hline \rule{1cm}{0.4pt} \end{array}$

10) $\begin{array}{r} 8,921.45 \\ -5,214.25 \\ \hline \rule{1cm}{0.4pt} \end{array}$

3) $\begin{array}{r} 582.54 \\ -321.45 \\ \hline \rule{1cm}{0.4pt} \end{array}$

7) $\begin{array}{r} 254.35 \\ +123.89 \\ \hline \rule{1cm}{0.4pt} \end{array}$

11) $\begin{array}{r} 2,321.25 \\ +1,984.99 \\ \hline \rule{1cm}{0.4pt} \end{array}$

4) $\begin{array}{r} 1,254 \\ +852.98 \\ \hline \rule{1cm}{0.4pt} \end{array}$

8) $\begin{array}{r} 3,257.5 \\ +1,245.2 \\ \hline \rule{1cm}{0.4pt} \end{array}$

12) $\begin{array}{r} 9,914.09 \\ -6,621.12 \\ \hline \rule{1cm}{0.4pt} \end{array}$

✎ Find the missing number.

13) $362.5 + \underline{} = 985.3$

14) $3,856 - \underline{} = 2,009.5$

15) $\underline{} - 985.1 = 1,450.9$

16) $2,785 - 1,234.12 = \underline{}$

17) $999.9 + \underline{} = 1,234.6$

18) $5,758.8 - 3,758.85 = \underline{}$

WWW.MathNotion.Com

Whole Number Multiplication and Division

✏ **Calculate each product.**

1) 35 × 43

2) 53.2 × 12.5

3) 37.2 × 16

4) 27.5 × 26

5) 158.8 × 15.4

6) 143.2 × 15.5

✏ **Find the missing quotient.**

7) $600 \div 1.5 =$ _____

8) $780 \div 39 =$ _____

9) $390 \div 1.3 =$ _____

10) $900 \div 0.9 =$ _____

11) $156 \div 40 =$ _____

12) $112 \div 1.6 =$ _____

13) $660 \div 2.2 =$ _____

14) $400 \div 0.8 =$ _____

15) $2{,}040 \div 25.5 =$ _____

16) $9{,}360 \div 31.2 =$ _____

✏ **Calculate each problem.**

17) $560 \div 7 = N$, $N =$ ___

18) $315 \div 4.5 = N$, $N =$ ___

19) $N \div 9 = 65$, $N =$ ___

20) $24.6 \times N = 147.6$, $N =$ ___

21) $985 \div N = 1{,}970$, $N =$ ___

22) $N \times 3.5 = 147$, $N =$ ___

Rounding and Estimates

🖎 **Estimate the sum by rounding each number to the nearest ten.**

1) $19 + 23 =$ _____

2) $72 + 31 =$ _____

3) $48 + 63 =$ _____

4) $44 + 86 =$ _____

5) $169 + 212 =$ _____

6) $650 + 323 =$ _____

7) $598 + 575 =$ _____

8) $1,586 + 3,355 =$ _____

🖎 **Estimate the product by rounding each number to the nearest ten.**

9) $37 \times 43 =$ _____

10) $12 \times 31 =$ _____

11) $48 \times 54 =$ _____

12) $17 \times 33 =$ _____

13) $68 \times 27 =$ _____

14) $91 \times 21 =$ _____

15) $86 \times 37 =$ _____

16) $96 \times 42 =$ _____

🖎 **Estimate the sum or product by rounding each number to the nearest ten.**

17) $\begin{array}{r} 28 \\ \times\ 16 \\ \hline \end{array}$

18) $\begin{array}{r} 72 \\ \times\ 22 \\ \hline \end{array}$

19) $\begin{array}{r} 85 \\ +\ 64 \\ \hline \end{array}$

20) $\begin{array}{r} 43 \\ +91 \\ \hline \end{array}$

21) $\begin{array}{r} 64 \\ \times\ 39 \\ \hline \end{array}$

22) $\begin{array}{r} 99 \\ +\ 54 \\ \hline \end{array}$

Adding and Subtracting Integers

✏ **Find each sum.**

1) $15 + (-35) =$

2) $(-28) + (-29) =$

3) $19 + (-27) =$

4) $57 + (-64) =$

5) $(-14) + (-19) + 64 =$

6) $54 + (-36) + 19 =$

7) $46 + (-30) + (-33) + 29 =$

8) $(-40) + (-70) + 28 + 55 =$

9) $60 + (-65) + (83 - 72) =$

10) $49 + (-55) + (90 - 67) =$

✏ **Find each difference.**

11) $(-32) - (-7) =$

12) $40 - (-12) =$

13) $(-60) - 56 =$

14) $27 - (-17) =$

15) $58 - (76 - 29) =$

16) $19 - (-14) - (-22) =$

17) $(39 + 15) - (-46) =$

18) $49 - 17 - (-13) =$

19) $85 - 45 - (-18) =$

20) $78 - (-35) - (-63) =$

21) $89 - (-11) - (-26) =$

22) $(19 - 50) - (-95) =$

23) $46 - 49 - (-87) =$

24) $120 - (98 + 24) - (-38) =$

25) $112 - (-102) + (-81) =$

26) $108 - (-42) + (-89) =$

Multiplying and Dividing Integers

✎ **Find each product.**

1) $(-7) \times (-9) =$

2) $(-5) \times 6 =$

3) $10 \times (-15) =$

4) $(-9) \times (-25) =$

5) $(-7) \times (-12) \times 13 =$

6) $(15 - 4) \times (-11) =$

7) $25 \times (-4) \times (-5) =$

8) $(85 + 10) \times (-11) =$

9) $12 \times (-19 + 12) \times 5 =$

10) $(-15) \times (-18) \times (-20) =$

✎ **Find each quotient.**

11) $85 \div (-5) =$

12) $(-90) \div (-15) =$

13) $(-121) \div (-11) =$

14) $99 \div (-33) =$

15) $(-114) \div 2 =$

16) $(-208) \div (-16) =$

17) $198 \div (-11) =$

18) $(-364) \div (-14) =$

19) $255 \div (-15) =$

20) $(-378) \div (18) =$

21) $(-184) \div (-8) =$

22) $-437 \div (-23) =$

23) $(-570) \div (-19) =$

24) $480 \div (-32) =$

25) $(-546) \div (-21) =$

26) $(486) \div (-54) =$

Order of Operations

✏️ **Evaluate each expression.**

1) $7 + (5 \times 8) =$

2) $16 - (6 \times 9) =$

3) $(17 \times 5) + 12 =$

4) $(24 - 12) - (11 \times 4) =$

5) $35 + (18 \div 3) =$

6) $(27 \times 3) \div 3 =$

7) $(88 \div 4) \times (-5) =$

8) $(9 \times 9) + (86 - 52) =$

9) $78 + (5 \times 12) + 14 =$

10) $(60 \times 4) \div (4 + 2) =$

11) $(-15) + (14 \times 4) + 18 =$

12) $(14 \times 5) - (56 \div 7) =$

13) $(7 \times 9 \div 3) - (32 + 21) =$

14) $(45 + 11 - 14) \times 2 - 15 =$

15) $(40 - 18 + 20) \times (75 \div 3) =$

16) $75 + (54 - (45 \div 9)) =$

17) $(12 + 15 - 24) + (44 \div 4) =$

18) $(78 - 19) + (27 - 10 + 7) =$

19) $(18 \times 3) + (17 \times 9) - 52 =$

20) $65 + 17 - (45 \times 2) + 40 =$

Ordering Integers and Numbers

✎ **Order each set of integers from least to greatest.**

1) $17, -15, -8, 0, 9$ ___, ___, ___, ___, ___, ___

2) $-14, -26, 17, 42, 39$ ___, ___, ___, ___, ___, ___

3) $32, -15, -69, 41, -80$ ___, ___, ___, ___, ___, ___

4) $-49, -65, 35, -21, 68$ ___, ___, ___, ___, ___, ___

5) $69, -32, 10, -45, 24$ ___, ___, ___, ___, ___, ___

6) $108, 76, -59, 87, -78$ ___, ___, ___, ___, ___, ___

✎ **Order each set of integers from greatest to least.**

7) $62, 98, -7, -19, -1$ ___, ___, ___, ___, ___, ___

8) $34, 35, -24, -46, 56$ ___, ___, ___, ___, ___, ___

9) $35, -96, -58, 17, -34$ ___, ___, ___, ___, ___, ___

10) $37, 12, -26, -13, 52$ ___, ___, ___, ___, ___, ___

11) $-12, 66, -18, -28, 54$ ___, ___, ___, ___, ___, ___

12) $-100, -85, -30, 5, 9$ ___, ___, ___, ___, ___, ___

Integers and Absolute Value

✍ **Write absolute value of each number.**

1) $|-19| =$

2) $|-32| =$

3) $|-50| =$

4) $|31| =$

5) $|57| =$

6) $|-76| =$

7) $|42| =$

8) $|101| =$

9) $|28| =$

10) $|-49| =$

11) $|-13|$

12) $|78| =$

13) $|100| =$

14) $|0| =$

15) $|-105| =$

16) $|-77| =$

17) $88 =$

18) $|-29| =$

19) $|112| =$

20) $|-120| =$

✍ **Evaluate the value.**

21) $|-5| - \dfrac{|-40|}{8} =$

22) $18 - |4 - 19| - |-15| =$

23) $\dfrac{|-72|}{9} \times |-9| =$

24) $\dfrac{|6 \times (-8)|}{3} \times \dfrac{|-21|}{7} =$

25) $|5 \times (-9)| + \dfrac{|-110|}{11} =$

26) $\dfrac{|-96|}{12} \times \dfrac{|-27|}{9} =$

27) $|-19 + 12| \times \dfrac{|-12 \times 13|}{7}$

28) $\dfrac{|-19 \times 6|}{3} \times |-11| =$

Factoring Numbers

✏️ **List all positive factors of each number.**

1) 6

2) 21

3) 28

4) 26

5) 46

6) 45

7) 48

8) 50

9) 52

10) 63

11) 70

12) 72

13) 78

14) 80

15) 82

16) 88

17) 90

18) 93

19) 95

20) 96

21) 98

22) 102

23) 124

24) 125

Greatest Common Factor

✏ **Find the GCF for each number pair.**

1) 6, 2

2) 8, 4

3) 5, 3

4) 6, 4

5) 7, 5

6) 8, 18

7) 14, 21

8) 6, 14

9) 9, 15

10) 4, 18

11) 14, 18

12) 25, 30

13) 27, 45

14) 36, 18

15) 9, 12

16) 11, 8

17) 28, 21

18) 56, 72

19) 34, 51

20) 6, 18, 27

21) 2, 9, 8

22) 10, 12, 24

23) 5, 14, 21

24) 72, 9, 18

Least Common Multiple

✎ **Find the LCM for each number pair.**

1) 6, 5

2) 8, 18

3) 9, 15

4) 15, 20

5) 20, 25

6) 22, 33

7) 6, 28

8) 8, 14

9) 21, 28

10) 14, 28

11) 9, 30

12) 7, 12

13) 12, 36

14) 9, 54

15) 42, 21

16) 40, 16

17) 12, 42

18) 13, 11

19) 32, 72

20) 15, 27

21) 24, 44

22) 8, 12, 42

23) 2, 6, 11

24) 15, 25, 30

Pre-Algebra Workbook

Answers of Worksheets – Chapter 1

Whole Number Addition and Subtraction

1) 2,877
2) 2,404
3) 261.09
4) 2,106.98
5) 1,984.35
6) 1,846.41
7) 378.24
8) 4,502.7
9) 6,426.1
10) 3,707.2
11) 4,306.24
12) 3,292.97
13) 622.8
14) 1,846.5
15) 2,436
16) 1,550.88
17) 234.7
18) 1,999.95

Whole Number Multiplication and Division

1) 1,505
2) 665
3) 595.2
4) 715
5) 2,445.52
6) 2,219.6
7) 400
8) 20
9) 300
10) 1,000
11) 3.9
12) 70
13) 300
14) 500
15) 80
16) 300
17) 80
18) 70
19) 585
20) 6
21) 2
22) 42

Rounding and Estimates

1) 40
2) 100
3) 110
4) 130
5) 380
6) 970
7) 1,180
8) 4,950
9) 1,600
10) 300
11) 2,500
12) 600
13) 2,100
14) 1,800
15) 3,600
16) 4,200
17) 600
18) 1,400
19) 150
20) 130
21) 2,400
22) 150

Adding and Subtracting Integers

1) −20
2) −57
3) −8
4) −7
5) 31
6) 37
7) 12
8) −27
9) 6
10) 17
11) −25
12) 52
13) −116
14) 44
15) 11
16) 55
17) 100
18) 45
19) 58
20) 176
21) 126
22) 64
23) 84
24) 36
25) 133
26) 61

Multiplying and Dividing Integers

1) 63
2) −30
3) −150
4) 225
5) 1,092
6) −121
7) 500
8) −1,045
9) −420
10) −5,400
11) −17
12) 6
13) 11
14) −3
15) −57
16) 13
17) −18
18) 26
19) −17
20) −21
21) 23
22) 19
23) 30
24) −15
25) 26
26) −9

Order of Operations

1) 47
2) −38
3) 97
4) −32
5) 41
6) 27
7) −110
8) 115
9) 152
10) 40
11) 59
12) 62
13) −32
14) 69
15) 1,050
16) 124
17) 14
18) 83
19) 155
20) 32

Ordering Integers and Numbers

1) −15, −8, 0, 9, 17
2) −26, −14, 17, 39, 42
3) −80, −69, −15, 32, 41
4) −65, −49, −21, 35, 68
5) −45, −32, 10, 24, 69
6) −78, −59, 76, 87, 108
7) 98, 62, −1, −7, −19
8) 56, 35, 34, −24, −46
9) 35, 17, −34, −58, −96
10) 52, 37, 12, −13, −26
11) 66, 54, −12, −18, −28
12) 9, 5, −30, −85, −100

Integers and Absolute Value

1) 19
2) 32
3) 50
4) 31
5) 57
6) 76
7) 42
8) 101
9) 28
10) 49
11) 13
12) 78
13) 100
14) 0
15) 105
16) 77
17) 88
18) 29
19) 112
20) 120
21) 0
22) -12
23) 72
24) 48
25) 55
26) 24
27) 156
28) 418

Factoring Numbers

1) 1, 2, 3, 6
2) 1, 3, 7, 21
3) 1, 2, 4, 7, 14, 28
4) 1, 2, 13, 26
5) 1, 2, 23, 46
6) 1, 3, 5, 9, 15, 45

Pre-Algebra Workbook

7) 1, 2, 3, 4, 6, 8, 12, 16, 24, 48
8) 1, 2, 5, 10, 25, 50
9) 1, 2, 4, 5, 13, 26, 52
10) 1, 3, 7, 9, 21, 63
11) 1, 2, 5, 7, 10, 14, 35, 70
12) 1, 2, 3, 4, 6, 8, 9, 12, 18, 24, 36, 72
13) 1, 2, 3, 6, 13, 26, 39, 78
14) 1, 2, 4, 5, 8, 10, 16, 20, 40, 80
15) 1, 2, 41, 82
16) 1, 2, 4, 8, 11, 22, 44, 88
17) 1, 2, 3, 5, 6, 9, 10, 15, 18, 30, 45, 90
18) 1, 3, 31, 93
19) 1, 5, 19, 95
20) 1, 2, 3, 4, 6, 8, 12, 16, 24, 32, 48, 96
21) 1, 2, 7, 14, 49, 98
22) 1, 2, 3, 6, 17, 34, 51, 102
23) 1, 2, 4, 31, 62, 124
24) 1, 5, 25, 125

Greatest Common Factor

1) 2
2) 4
3) 1
4) 2
5) 1
6) 2
7) 7
8) 2
9) 3
10) 2
11) 2
12) 5
13) 9
14) 18
15) 3
16) 1
17) 7
18) 8
19) 17
20) 3
21) 1
22) 2
23) 1
24) 9

Least Common Multiple

1) 30
2) 72
3) 45
4) 60
5) 100
6) 66
7) 84
8) 56
9) 84
10) 28
11) 90
12) 84
13) 36
14) 54
15) 42
16) 80
17) 84
18) 143
19) 288
20) 135
21) 264
22) 168
23) 66
24) 150

Chapter 2: Fractions and Decimals

Topics that you will practice in this chapter:

- ✓ Simplifying Fractions
- ✓ Adding and Subtracting Fractions
- ✓ Multiplying and Dividing Fractions
- ✓ Adding and Subtract Mixed Numbers
- ✓ Multiplying and Dividing Mixed Numbers
- ✓ Adding and Subtracting Decimals
- ✓ Multiplying and Dividing Decimals
- ✓ Comparing Decimals
- ✓ Rounding Decimals

"A Man is like a fraction whose numerator is what he is and whose denominator is what he thinks of himself. The larger the denominator, the smaller the fraction." –Tolstoy

Simplifying Fractions

✎ **Simplify each fraction to its lowest terms.**

1) $\frac{8}{16} =$

2) $\frac{28}{35} =$

3) $\frac{27}{36} =$

4) $\frac{70}{140} =$

5) $\frac{13}{52} =$

6) $\frac{38}{57} =$

7) $\frac{64}{80} =$

8) $\frac{21}{84} =$

9) $\frac{85}{170} =$

10) $\frac{120}{168} =$

11) $\frac{31}{124} =$

12) $\frac{48}{96} =$

13) $\frac{98}{112} =$

14) $\frac{99}{110} =$

15) $\frac{51}{153} =$

16) $\frac{40}{112} =$

17) $\frac{90}{225} =$

18) $\frac{44}{297} =$

19) $\frac{54}{279} =$

20) $\frac{320}{720} =$

21) $\frac{70}{560} =$

✎ **Find the answer for each problem.**

22) Which of the following fractions equal to $\frac{3}{7}$? ____

    A. $\frac{24}{63}$ B. $\frac{51}{109}$ C. $\frac{51}{119}$ D. $\frac{240}{630}$

23) Which of the following fractions equal to $\frac{7}{8}$? ____

    A. $\frac{182}{208}$ B. $\frac{175}{208}$ C. $\frac{182}{216}$ D. $\frac{49}{64}$

24) Which of the following fractions equal to $\frac{2}{9}$? ____

    A. $\frac{64}{126}$ B. $\frac{46}{207}$ C. $\frac{48}{207}$ D. $\frac{56}{208}$

Adding and Subtracting Fractions

🖋 **Find the sum.**

1) $\frac{5x}{8} + \frac{3x}{8} =$

2) $\frac{x}{2} + \frac{x}{7} =$

3) $\frac{y}{3} + \frac{y}{4} =$

4) $\frac{3x}{8} + \frac{2x}{5} =$

5) $\frac{xy}{5} + \frac{2xy}{7} =$

6) $\frac{2x}{9} + \frac{4x}{9} =$

7) $\frac{a}{4} + \frac{2a}{3} =$

8) $\frac{2}{x} + \frac{4}{x} =$

9) $\frac{1}{a} + \frac{2}{b} =$

10) $\frac{3b}{5} + \frac{2b}{7} =$

11) $\frac{a}{y} + \frac{3a}{y} =$

12) $\frac{3}{x} + \frac{1}{2x} =$

🖋 **Find the difference.**

13) $\frac{x}{3} - \frac{x}{6} =$

14) $\frac{2x}{5} - \frac{3x}{8} =$

15) $\frac{x}{7} - \frac{y}{7} =$

16) $\frac{2x}{7} - \frac{x}{6} =$

17) $\frac{5a}{9} - \frac{2a}{5} =$

18) $\frac{2ab}{3} - \frac{ab}{6} =$

19) $\frac{1}{x} - \frac{1}{3x} =$

20) $\frac{4}{y} - \frac{3}{4y} =$

21) $\frac{5}{x} - \frac{2y}{xy} =$

22) $\frac{8}{ab} - \frac{5}{3ab} =$

23) $\frac{2a}{y} - \frac{a}{3y} =$

24) $\frac{5}{b} - \frac{2}{3b} =$

25) $\frac{2a}{b} - \frac{a}{b} =$

26) $\frac{3}{a} - \frac{2}{b} =$

27) $\frac{4a}{b} - \frac{2a}{3b} =$

28) $\frac{6}{xy} - \frac{7}{2xy} =$

29) $\frac{2}{a} - \frac{1}{4a} =$

30) $\frac{2a}{3b} - \frac{4a}{9b} =$

Multiplying and Dividing Fractions

✏️ **Find the value of each expression in lowest terms.**

1) $\dfrac{3}{a} \times \dfrac{5}{3} =$

2) $\dfrac{2}{3b} \times \dfrac{9}{2} =$

3) $\dfrac{a}{15} \times \dfrac{5}{2a} =$

4) $\dfrac{x}{3a} \times \dfrac{9a}{6x} =$

5) $\dfrac{x}{12} \times \dfrac{6}{y} =$

6) $\dfrac{7}{x} \times \dfrac{x}{14} =$

7) $\dfrac{10}{3a} \times \dfrac{6}{20} =$

8) $\dfrac{4a}{b} \times \dfrac{2b}{5} =$

9) $\dfrac{2ab}{7} \times \dfrac{14}{6ab} =$

10) $\dfrac{4a}{5b} \times \dfrac{15}{2a} =$

11) $\dfrac{ab}{21} \times \dfrac{7}{a} =$

12) $\dfrac{a}{cd} \times \dfrac{2bc}{a} =$

✏️ **Find the value of each expression in lowest terms.**

13) $\dfrac{a}{2} \div \dfrac{a}{4} =$

14) $\dfrac{b}{3} \div \dfrac{b}{9} =$

15) $\dfrac{a}{b} \div \dfrac{3}{b} =$

16) $\dfrac{2a}{15} \div \dfrac{4a}{5} =$

17) $\dfrac{1}{a} \div \dfrac{b}{3a} =$

18) $\dfrac{4a}{3b} \div \dfrac{a}{2b} =$

19) $\dfrac{a}{8} \div \dfrac{3a}{16} =$

20) $\dfrac{3b}{20} \div \dfrac{6b}{15a} =$

21) $\dfrac{x}{12y} \div \dfrac{2x}{9y} =$

22) $\dfrac{25}{x} \div \dfrac{50}{2x} =$

23) $\dfrac{16}{5ab} \div \dfrac{32}{ab} =$

24) $\dfrac{7a}{b} \div \dfrac{8a}{b} =$

25) $\dfrac{5}{x} \div \dfrac{3y}{x} =$

26) $\dfrac{2a}{21} \div \dfrac{a}{14} =$

27) $\dfrac{ab}{x} \div \dfrac{a}{x} =$

28) $\dfrac{6}{a} \div \dfrac{3b}{2a} =$

29) $\dfrac{9}{16a} \div \dfrac{3}{8ab} =$

30) $\dfrac{24}{xy} \div \dfrac{12}{y} =$

Adding and Subtracting Mixed Numbers

🪶 **Find the sum.**

1) $3\frac{5}{6} + 2\frac{1}{3} =$

2) $4\frac{2}{5} + 1\frac{1}{5} =$

3) $5\frac{1}{8} + 6\frac{3}{4} =$

4) $2\frac{2}{3} + 3\frac{1}{2} =$

5) $3\frac{4}{5} + 3\frac{2}{15} =$

6) $8\frac{1}{16} + 3\frac{3}{8} =$

7) $4\frac{3}{5} + 4\frac{1}{6} =$

8) $7\frac{3}{4} + 3\frac{5}{6} =$

9) $8\frac{5}{6} + 2\frac{2}{7} =$

10) $11\frac{3}{16} + 3\frac{5}{24} =$

🪶 **Find the difference.**

11) $3\frac{3}{4} - 2\frac{1}{4} =$

12) $5\frac{1}{7} - 3\frac{1}{7} =$

13) $4\frac{1}{3} - 1\frac{1}{9} =$

14) $7\frac{1}{6} - 3\frac{1}{12} =$

15) $6\frac{1}{3} - 2\frac{5}{18} =$

16) $8\frac{1}{4} - 5\frac{1}{8} =$

17) $9\frac{1}{2} - 6\frac{1}{5} =$

18) $11\frac{7}{15} - 8\frac{1}{30} =$

19) $12\frac{3}{5} - 7\frac{2}{7} =$

20) $18\frac{1}{8} - 14\frac{3}{16} =$

21) $12\frac{2}{3} - 11\frac{7}{15} =$

22) $3\frac{1}{5} - 1\frac{1}{2} =$

23) $14\frac{3}{5} - 6\frac{4}{5} =$

24) $17\frac{1}{4} - 14\frac{8}{9} =$

25) $24\frac{3}{9} - 15\frac{1}{18} =$

26) $28\frac{3}{7} - 19\frac{5}{6} =$

Multiplying and Dividing Mixed Numbers

✍ **Find the product.**

1) $2\frac{1}{3} \times 4\frac{1}{2} =$

2) $4\frac{1}{5} \times 2\frac{1}{3} =$

3) $7\frac{2}{3} \times 3\frac{3}{5} =$

4) $9\frac{2}{7} \times 3\frac{1}{8} =$

5) $5\frac{4}{11} \times 4\frac{1}{3} =$

6) $7\frac{3}{8} \times 5\frac{4}{9} =$

7) $9\frac{2}{3} \times 11\frac{5}{6} =$

8) $8\frac{3}{5} \times 7\frac{4}{9} =$

9) $5\frac{1}{9} \times 9\frac{5}{8} =$

10) $10\frac{2}{7} \times 2\frac{5}{8} =$

✍ **Find the quotient.**

11) $2\frac{1}{8} \div 1\frac{3}{8} =$

12) $4\frac{1}{6} \div 2\frac{1}{3} =$

13) $7\frac{1}{3} \div 3\frac{3}{4} =$

14) $4\frac{5}{8} \div 1\frac{1}{2} =$

15) $6\frac{5}{12} \div 4\frac{1}{6} =$

16) $5\frac{7}{18} \div 5\frac{1}{6} =$

17) $6\frac{5}{21} \div 2\frac{3}{7} =$

18) $8\frac{1}{7} \div 8\frac{1}{14} =$

19) $10\frac{1}{4} \div 3\frac{2}{5} =$

20) $15\frac{1}{3} \div 5\frac{2}{9} =$

21) $12\frac{1}{3} \div 6\frac{1}{2} =$

22) $18\frac{1}{9} \div 18\frac{1}{6} =$

23) $10\frac{3}{4} \div 5\frac{2}{5} =$

24) $11\frac{1}{3} \div 8\frac{4}{5} =$

25) $9\frac{1}{6} \div 3\frac{2}{7} =$

26) $7\frac{1}{3} \div 3\frac{7}{11} =$

Adding and Subtracting Decimals

✍ **Add and subtract decimals.**

1) 52.18 − 21.27

2) 49.34 + 25.24

3) 48.60 + 35.75

4) 65.84 − 35.49

5) 54.57 + 18.37

6) 90.45 − 28.75

7) 98.12 − 45.55

8) 48.99 + 57.67

9) 158.05 − 78.98

✍ **Find the missing number.**

10) ___ + 4.9 = 6.5

11) 5.15 + ___ = 6.43

12) 8.09 + ___ = 11.84

13) 8.88 − ___ = 6.78

14) ___ − 1.59 = 3.71

15) ___ − 19.98 = 8.17

16) 38.89 + ___ = 41.32

17) ___ − 35.99 = 1.80

18) ___ + 39.08 = 41.36

19) 98.98 + ___ = 123.68

Multiplying and Dividing Decimals

🖉 Find the product.

1) $0.6 \times 0.8 =$

2) $2.5 \times 0.9 =$

3) $0.87 \times 0.4 =$

4) $0.15 \times 0.75 =$

5) $0.95 \times 0.7 =$

6) $1.57 \times 0.9 =$

7) $5.85 \times 1.3 =$

8) $12.5 \times 4.5 =$

9) $19.8 \times 7.32 =$

10) $85.1 \times 1.5 =$

11) $79.5 \times 11.2 =$

12) $86.9 \times 21.5 =$

🖉 Find the quotient.

13) $3.25 \div 10 =$

14) $24.5 \div 100 =$

15) $3.9 \div 3 =$

16) $91.2 \div 0.6 =$

17) $29.2 \div 0.4 =$

18) $38.7 \div `9 =$

19) $297.8 \div 1,000 =$

20) $53.55 \div 0.7 =$

21) $345.45 \div 0.1 =$

22) $70.27 \div 0.25 =$

23) $28.968 \div 0.3 =$

24) $86.34 \div 0.06 =$

Comparing Decimals

✍ **Write the correct comparison symbol (>, < or =).**

1) 0.80 ☐ 0.080

2) 0.086 ☐ 0.86

3) 7.090 ☐ 7.09

4) 3.25 ☐ 3.06

5) 4.09 ☐ 0.490

6) 6.06 ☐ 6.6

7) 6.08 ☐ 6.080

8) 4.05 ☐ 4.2

9) 12.35 ☐ 12.198

10) 0.957 ☐ 0.0957

11) 25.24 ☐ 25.240

12) 0.742 ☐ 0.752

13) 14.09 ☐ 14.10

14) 17.45 ☐ 17.154

15) 11.44 ☐ 11.439

16) 15.41 ☐ 15.410

17) 21.43 ☐ 21.043

18) 8.098 ☐ 8.90

19) 16.044 ☐ 16.040

20) 32.35 ☐ 32.350

Rounding Decimals

🪶 Round each decimal to the nearest whole number.

1) 56.27   3) 18.32   5) 7.90

2) 5.9   4) 4.8   6) 57.7

🪶 Round each decimal to the nearest tenth.

7) 42.785   9) 96.586   11) 27.198

8) 15.224   10) 101.78   12) 96.87

🪶 Round each decimal to the nearest hundredth.

13) 9.648   15) 89.2882   17) 68.229

14) 27.819   16) 120.912   18) 85.642

🪶 Round each decimal to the nearest thousandth.

19) 19.88486   21) 145.9322   23) 189.0991

20) 46.72611   22) 210.1581   24) 121.76798

Answers of Worksheets – Chapter 2

Simplifying Fractions

1) $\frac{1}{2}$
2) $\frac{4}{5}$
3) $\frac{3}{4}$
4) $\frac{1}{2}$
5) $\frac{1}{4}$
6) $\frac{2}{3}$
7) $\frac{4}{5}$
8) $\frac{1}{4}$
9) $\frac{1}{2}$
10) $\frac{5}{7}$
11) $\frac{1}{4}$
12) $\frac{1}{2}$
13) $\frac{7}{8}$
14) $\frac{9}{10}$
15) $\frac{1}{3}$
16) $\frac{5}{14}$
17) $\frac{2}{5}$
18) $\frac{4}{27}$
19) $\frac{6}{31}$
20) $\frac{4}{9}$
21) $\frac{1}{8}$
22) C
23) A
24) B

Adding and Subtracting Fractions

1) $\frac{8x}{8} = x$
2) $\frac{9x}{14}$
3) $\frac{7x}{12}$
4) $\frac{31x}{40}$
5) $\frac{17xy}{35}$
6) $\frac{2x}{3}$
7) $\frac{11a}{12}$
8) $\frac{6}{x}$
9) $\frac{a+2b}{ab}$
10) $\frac{31b}{35}$
11) $\frac{4a}{y}$
12) $\frac{7}{2x}$
13) $\frac{x}{6}$
14) $\frac{x}{40}$
15) $\frac{x-y}{7}$
16) $\frac{5x}{42}$
17) $\frac{7a}{45}$
18) $\frac{ab}{2}$
19) $\frac{2}{3x}$
20) $\frac{13}{4y}$
21) $\frac{3}{x}$
22) $\frac{19}{3ab}$
23) $\frac{5a}{3y}$
24) $\frac{13}{3b}$
25) $\frac{a}{b}$
26) $\frac{3b-2a}{ab}$
27) $\frac{10a}{3b}$
28) $\frac{5}{2xy}$
29) $\frac{7}{4a}$
30) $\frac{2a}{9b}$

Multiplying and Dividing Fractions

1) $\frac{5}{a}$
2) $\frac{3}{b}$
3) $\frac{1}{6}$
4) $\frac{1}{2}$
5) $\frac{x}{2y}$
6) $\frac{1}{2}$
7) $\frac{1}{a}$
8) $\frac{8a}{5}$
9) $\frac{2}{3}$

10) $\dfrac{6}{b}$

11) $\dfrac{b}{3}$

12) $\dfrac{2b}{d}$

13) 2

14) 3

15) $\dfrac{a}{3}$

16) $\dfrac{1}{6}$

17) $\dfrac{3}{b}$

18) $\dfrac{8}{3}$

19) $\dfrac{2}{3}$

20) $\dfrac{3a}{8}$

21) $\dfrac{3}{8}$

22) 1

23) $\dfrac{1}{10}$

24) $\dfrac{7}{8}$

25) $\dfrac{5}{3y}$

26) $\dfrac{4}{3}$

27) b

28) $\dfrac{4}{b}$

29) $\dfrac{3b}{2}$

30) $\dfrac{2}{x}$

Adding and Subtracting Mixed Numbers

1) $6\dfrac{1}{6}$

2) $5\dfrac{3}{5}$

3) $11\dfrac{7}{8}$

4) $6\dfrac{1}{6}$

5) $6\dfrac{14}{15}$

6) $11\dfrac{7}{16}$

7) $8\dfrac{23}{30}$

8) $11\dfrac{7}{12}$

9) $11\dfrac{5}{42}$

10) $14\dfrac{19}{48}$

11) $1\dfrac{1}{2}$

12) 2

13) $3\dfrac{2}{9}$

14) $4\dfrac{1}{12}$

15) $4\dfrac{1}{18}$

16) $3\dfrac{1}{8}$

17) $3\dfrac{3}{10}$

18) $3\dfrac{13}{30}$

19) $5\dfrac{11}{35}$

20) $3\dfrac{15}{16}$

21) $1\dfrac{1}{5}$

22) $1\dfrac{7}{10}$

23) $7\dfrac{4}{5}$

24) $2\dfrac{13}{36}$

25) $9\dfrac{5}{18}$

26) $8\dfrac{25}{42}$

Multiplying and Dividing Mixed Numbers

1) $10\dfrac{1}{2}$

2) $9\dfrac{4}{5}$

3) $27\dfrac{3}{5}$

4) $29\dfrac{1}{56}$

5) $23\dfrac{8}{33}$

6) $40\dfrac{11}{72}$

7) $144\dfrac{7}{18}$

8) $64\dfrac{1}{45}$

9) $49\dfrac{7}{36}$

10) 27

11) $1\dfrac{6}{11}$

12) $1\dfrac{11}{14}$

13) $1\dfrac{43}{45}$

14) $3\dfrac{1}{12}$

15) $1\dfrac{27}{50}$

16) $1\dfrac{4}{93}$

17) $2\dfrac{29}{51}$

18) $1\dfrac{1}{113}$

19) $3\dfrac{1}{68}$

20) $2\dfrac{44}{47}$

21) $1\dfrac{35}{39}$

22) $\frac{326}{327}$ 24) $1\frac{19}{66}$ 26) $2\frac{1}{60}$

23) $1\frac{107}{108}$ 25) $2\frac{109}{138}$

Adding and Subtracting Decimals

1) 30.91   6) 61.7   11) 1.28   16) 2.43
2) 74.58   7) 52.57   12) 3.75   17) 37.79
3) 84.35   8) 106.66   13) 2.1   18) 2.28
4) 30.35   9) 79.07   14) 5.3   19) 24.7
5) 72.94   10) 1.6   15) 28.15

Multiplying and Dividing Decimals

1) 0.48   7) 7.605   13) 0.325   19) 0.2978
2) 2.25   8) 56.25   14) 0.245   20) 76.5
3) 0.348   9) 144.936   15) 1.3   21) 3,454.5
4) 0.1125   10) 127.65   16) 152   22) 281.08
5) 0.665   11) 890.4   17) 73   23) 96.56
6) 1.413   12) 1,868.35   18) 4.3   24) 1,439

Comparing Decimals

1) >   6) <   11) =   16) =
2) <   7) =   12) <   17) >
3) =   8) <   13) <   18) <
4) >   9) >   14) >   19) >
5) >   10) >   15) >   20) =

Rounding Decimals

1) 56   9) 96.6   17) 68.23
2) 6   10) 101.8   18) 85.64
3) 18   11) 27.2   19) 19.885
4) 5   12) 96.9   20) 46.726
5) 8   13) 9.65   21) 145.932
6) 58   14) 27.82   22) 210.158
7) 42.8   15) 89.29   23) 189.099
8) 15.2   16) 120.91   24) 121.768

Chapter 3:
Proportions, Ratios, and Percent

Topics that you will practice in this chapter:

- ✓ Simplifying Ratios
- ✓ Proportional Ratios
- ✓ Similarity and Ratios
- ✓ Ratio and Rates Word Problems
- ✓ Percentage Calculations
- ✓ Percent Problems
- ✓ Discount, Tax and Tip
- ✓ Percent of Change
- ✓ Simple Interest

Without mathematics, there's nothing you can do. Everything around you is mathematics. Everything around you is numbers." – Shakuntala Devi

Simplifying Ratios

✎ **Reduce each ratio.**

1) $15:20 = $ ___ : ___

2) $9:90 = $ ___ : ___

3) $24:42 = $ ___ : ___

4) $7:21 = $ ___ : ___

5) $11:110$ ___ : ___

6) $8:64 = $ ___ : ___

7) $18:72 = $ ___ : ___

8) $10:25 = $ ___ : ___

9) $7:42 = $ ___ : ___

10) $49:63 = $ ___ : ___

11) $12:18$ ___ : ___

12) $35:10$ ___ : ___

13) $150:15$ ___ : ___

14) $2.4:3.2$ ___ : ___

15) $7:56 = $ ___ : ___

16) $45:63$ ___ : ___

17) $77:99$ ___ : ___

18) $39:13$ ___ : ___

19) $15:45$ ___ : ___

20) $84:12$ ___ : ___

21) $25:5$ ___ : ___

22) $70:56$ ___ : ___

23) $70:140$ ___ : ___

24) $1.2:36$ ___ : ___

✎ **Write each ratio as a fraction in simplest form.**

25) $7:14 =$

26) $27:45 =$

27) $24:56 =$

28) $16:48 =$

29) $22:66 =$

30) $21:98 =$

31) $34:68 =$

32) $6:30 =$

33) $35:84 =$

34) $12:54 =$

35) $88:104 =$

36) $36:81 =$

37) $1.5:18 =$

38) $4.5:16.5 =$

39) $5:75 =$

40) $3.1:12.4 =$

41) $1.6:6.4 =$

42) $0.25:1.25 =$

43) $8.8:16.4 =$

44) $0.75:6.75 =$

45) $1.8:3 =$

Proportional Ratios

✎ **Fill in the blanks; Calculate each proportion.**

1) $3:8 = \_\_ : 32$

2) $1:2 = 45: \_\_$

3) $1:11 = \_\_ : 55$

4) $9:12 = 18: \_\_$

5) $9:7 = 81: \_\_$

6) $2:8 = \_\_ : 56$

7) $2.3:1.2 = \_\_ : 12$

8) $0.5:2 = \_\_ : 32$

9) $1.6:2 = \_\_ : 60$

10) $2.5:4.5 = \_\_ : 90$

11) $3.8:7.1 = 7.6: \_\_$

12) $5.5:6 = 16.5: \_\_$

✎ **State if each pair of ratios form a proportion.**

13) $\frac{5}{12}$ and $\frac{15}{36}$

14) $\frac{2}{4}$ and $\frac{18}{36}$

15) $\frac{7}{8}$ and $\frac{28}{32}$

16) $\frac{3}{8}$ and $\frac{27}{64}$

17) $\frac{1}{14}$ and $\frac{5}{65}$

18) $\frac{7}{11}$ and $\frac{70}{100}$

19) $\frac{12}{15}$ and $\frac{48}{60}$

20) $\frac{3}{17}$ and $\frac{36}{204}$

21) $\frac{1.2}{1.5}$ and $\frac{1.44}{22.5}$

22) $\frac{1.3}{1.1}$ and $\frac{3.9}{33}$

23) $\frac{0.7}{0.9}$ and $\frac{6.3}{8.1}$

24) $\frac{2.4}{3.2}$ and $\frac{48}{64}$

✎ **Calculate each proportion.**

25) $\frac{14}{16} = \frac{21}{x}, x = \_\_\_$

26) $\frac{3}{28} = \frac{42}{x}, x = \_\_\_$

27) $\frac{19}{5} = \frac{38}{x}, x = \_\_\_$

28) $\frac{3}{10} = \frac{x}{140}, x = \_\_\_$

29) $\frac{4}{9} = \frac{x}{108}, x = \_\_\_$

30) $\frac{7}{32} = \frac{21}{x}, x = \_\_\_$

31) $\frac{9}{8} = \frac{108}{x}, x = \_\_\_$

32) $\frac{12}{17} = \frac{48}{x}, x = \_\_\_$

33) $\frac{1.4}{5} = \frac{x}{30}, x = \_\_\_$

34) $\frac{1.6}{12} = \frac{x}{60}, x = \_\_\_$

35) $\frac{3.5}{15} = \frac{x}{315}, x = \_\_\_$

36) $\frac{4.7}{2.5} = \frac{x}{50}, x = \_\_\_$

Pre-Algebra Workbook

Similarity and Ratios

✍ **Each pair of figures is similar. Find the missing side.**

1) [Triangles: first with legs 24 and 18, hypotenuse ?; second with 12, 9, and 15]

2) [Triangles: first with sides 12.5, 7.5, base 5; second with sides 5, 3, base ?]

3) [Triangles: first with side 24, base 60; second with side 4, base ?]

4) [Trapezoids: first with sides ? and 7; second with 132 and 84]

✍ **Calculate.**

5) Two rectangles are similar. The first is 14 feet wide and 70 feet long. The second is 30 feet wide. What is the length of the second rectangle? _____

6) Two rectangles are similar. One is 3.2 meters by 15 meters. The longer side of the second rectangle is 42 meters. What is the other side of the second rectangle? _____

7) A building casts a shadow 24 ft long. At the same time a girl 10 ft tall casts a shadow 6 ft long. How tall is the building? _____

8) The scale of a map of Texas is 8 inches: 52 miles. If you measure the distance from Dallas to Martin County as 28.8 inches, approximately how far is Martin County from Dallas? _____

WWW.MathNotion.Com

Ratio and Rates Word Problems

✎ **Find the answer for each word problem.**

1) Mason has 32 red cards and 40 green cards. What is the ratio of Mason's red cards to his green cards? _____

2) In a party, 24 soft drinks are required for every 42 guests. If there are 378 guests, how many soft drinks is required? _____

3) In Mason's class, 54 of the students are tall and 30 are short. In Michael's class 126 students are tall and 70 students are short. Which class has a higher ratio of tall to short students? _____

4) The price of 4 apples at the Quick Market is $3.65. The price of 6 of the same apples at Walmart is $4.25. Which place is the better buy? _____

5) The bakers at a Bakery can make 90 bagels in 3 hours. How many bagels can they bake in 17 hours? What is that rate per hour? _____

6) You can buy 8 cans of green beans at a supermarket for $5.60. How much does it cost to buy 56 cans of green beans? _____

7) The ratio of boys to girls in a class is 4: 7. If there are 16 boys in the class, how many girls are in that class? _____

8) The ratio of red marbles to blue marbles in a bag is 3: 4. If there are 42 marbles in the bag, how many of the marbles are red? _____

Percentage Calculations

✎ **Calculate the given percent of each value.**

1) 3% of 60 = ____
2) 20% of 80 = ____
3) 25% of 80 = ____
4) 24% of 50 = ____
5) 18% of 150 = ____
6) 70% of 35 = ____

7) 15% of 28 = ____
8) 32% of 300 = ____
9) 54% of 80 = ____
10) 10% of 610 = ____
11) 35% of 520 = ____
12) 64% of 110 = ____

13) 44% of 200 = ____
14) 28% of 94 = ____
15) 30% of 85 = ____
16) 68% of 102 = ____
17) 45% of 160 = ____
18) 55% of 220 = ____

✎ **Calculate the percent of each given value.**

19) ____% of 18 = 9
20) ____% of 50 = 40
21) ____% of 140 = 7
22) ____% of 158 = 39.5
23) ____% of 75 = 9.375

24) ____% of 45 = 11.25
25) ____% of 90 = 22.5
26) ____% of 650 = 19.5
27) ____% of 480 = 24
28) ____% of 400 = 57.32

✎ **Calculate each percent problem.**

29) A Cinema has 132 seats. 92 seats were sold for the current movie. What percent of seats are empty? ____ %

30) There are 52 boys and 68 girls in a class. 55.00% of the students in the class take the bus to school. How many students do not take the bus to school? ____

Percent Problems

✎ Calculate each problem.

1) 30 is what percent of 60? ____%

2) 32 is what percent of 80? ____%

3) 72 is what percent of 45? ____%

4) 8 is what percent of 200? ____%

5) 9 is what percent of 600? ____%

6) 30 is what percent of 500? ____%

7) 70 is what percent of 350? ____%

8) 44 is what percent of 550? ____%

9) 270 is what percent of 900? ____%

10) 180 is what percent of 720? ___%

11) 37.5 is what percent of 75? ___%

12) 27.5 is what percent of 55? ___%

13) 60 is what percent of 750? ___%

14) 22.5 is what percent of 18? ___%

15) 36 is what percent of 24? ___%

16) 18 is what percent of 60? ___%

17) 140 is what percent of 280? ___%

18) 128 is what percent of 40? ___%

✎ Calculate each percent word problem.

19) There are 32 employees in a company. On a certain day, 24 were present. What percent showed up for work? _____%

20) A metal bar weighs 36 ounces. 40% of the bar is gold. How many ounces of gold are in the bar? _____

21) A crew is made up of 12 women; the rest are men. If 20% of the crew are women, how many people are in the crew? _____

22) There are 32 students in a class and 8 of them are girls. What percent are boys? _____%

23) The Royals softball team played 310 games and won 248 of them. What percent of the games did they lose? _____%

Discount, Tax and Tip

✍ Find the selling price of each item.

1) Original price of a computer: $450
   Tax: 8% Selling price: $_____

2) Original price of a laptop: $240
   Tax: 4% Selling price: $_____

3) Original price of a sofa: $900
   Tax: 12%   Selling price: $_____

4) Original price of a car: $10,400
   Tax: 2.5%   Selling price: $_____

5) Original price of a Table: $400
   Tax: 3% Selling price: $_____

6) Original price of a house: $360,000
   Tax: 2.8%   Selling price: $_____

7) Original price of a tablet: $150
   Discount: 24%   Selling price: $____

8) Original price of a chair: $180
   Discount: 20%   Selling price: $____

9) Original price of a book: $80
   Discount: 30%   Selling price: $____

10) Original price of a cellphone: $800
    Discount: 20%   Selling price: $____

11) Food bill: $56
    Tip:15%   Price: $_____

12) Food bill: $50
    Tipp: 10%   Price: $_____

13) Food bill: $94
    Tip: 25%   Price: $_____

14) Food bill: $48
    Tipp: 30%   Price: $_____

✍ Find the answer for each word problem.

15) Nicolas hired a moving company. The company charged $400 for its services, and Nicolas gives the movers a 20% tip. How much does Nicolas tip the movers? $_____

16) Mason has lunch at a restaurant and the cost of his meal is $80. Mason wants to leave a 20% tip. What is Mason's total bill including tip? $_____

17) The sales tax in Texas is 14.45% and an item costs $300. How much is the tax? $_____

18) The price of a table at Best Buy is $520. If the sales tax is 4%, what is the final price of the table including tax? $_____

Percent of Change

Find each percent of change.

1) From 300 to 600. ___ %
2) From 45 ft to 225 ft. ___ %
3) From $60 to $420. ___ %
4) From 30 cm to 120 cm. ___ %
5) From 10 to 30. ___ %
6) From 12 to 30. ___ %
7) From 140 to 210. ___ %
8) From 800 to 400. ___ %
9) From 85 to 51. ___ %
10) From 152 to 76. ___ %

Calculate each percent of change word problem.

11) Bob got a raise, and his hourly wage increased from $32 to $40. What is the percent increase? ____ %

12) The price of a pair of shoes increases from $70 to $112. What is the percent increase? ___ %

13) At a coffee shop, the price of a cup of coffee increased from $1.90 to $2.28. What is the percent increase in the cost of the coffee? _____ %

14) 30 cm are cut from a 120 cm board. What is the percent decrease in length? _____ %

15) In a class, the number of students has been increased from 54 to 81. What is the percent increase? _____ %

16) The price of gasoline rose from $22.4 to $25.76 in one month. By what percent did the gas price rise? _____ %

17) A shirt was originally priced at $19. It went on sale for $22.80. What was the percent that the shirt was discounted? _____ %

Simple Interest

🖋 **Determine the simple interest for these loans.**

1) $210 at 15% for 4 years. $ _____

2) $1,200 at 6% for 3 years. $ _____

3) $950 at 25% for 2 years. $ _____

4) $6,500 at 1.5% for 7 months. $ ____

5) $240 at 5% for 8 months. $ _____

6) $28,000 at 3.5% for 6 years. $ _____

7) $9,600 at 8% for 2 years. $ _____

8) $500 at 4.2% for 5 years. $ _____

9) $700 at 2.8 % for 6 months. $ ____

10) $9,000 at 1.6% for 4 years. $ _____

🖋 **Calculate each simple interest word problem.**

11) A new car, valued at $16,000, depreciates at 3.5% per year. What is the value of the car two year after purchase? $_____

12) Sara puts $9,000 into an investment yielding 8% annual simple interest; she left the money in for three years. How much interest does Sara get at the end of those three years? $_____

13) A bank is offering 12.5% simple interest on a savings account. If you deposit $32,400, how much interest will you earn in one years? $_____

14) $2,400 interest is earned on a principal of $10,000 at a simple interest rate of 12% interest per year. For how many years was the principal invested? _____

15) In how many years will $1,200 yield an interest of $384 at 8% simple interest? _____

16) Jim invested $5,000 in a bond at a yearly rate of 2.5%. He earned $375 in interest. How long was the money invested? _____

Answers of Worksheets – Chapter 3

Simplifying Ratios

1) 3:4
2) 1:10
3) 4:7
4) 1:3
5) 1:10
6) 1:8
7) 2:8
8) 2:5
9) 1:6
10) 7:9
11) 2:3
12) 7:2
13) 10:1
14) 3:4
15) 1:8
16) 5:7
17) 7:9
18) 3:1
19) 1:3
20) 7:1
21) 5:1
22) 5:4
23) 1:2
24) 1:30
25) $\frac{1}{2}$
26) $\frac{3}{5}$
27) $\frac{3}{7}$
28) $\frac{1}{3}$
29) $\frac{1}{3}$
30) $\frac{3}{14}$
31) $\frac{1}{2}$
32) $\frac{1}{5}$
33) $\frac{5}{12}$
34) $\frac{2}{9}$
35) $\frac{11}{13}$
36) $\frac{4}{9}$
37) $\frac{1}{12}$
38) $\frac{3}{11}$
39) $\frac{1}{15}$
40) $\frac{1}{4}$
41) $\frac{4}{3}$
42) $\frac{1}{5}$
43) $\frac{22}{41}$
44) $\frac{1}{9}$
45) $\frac{3}{5}$

Proportional Ratios

1) 12
2) 90
3) 5
4) 24
5) 63
6) 14
7) 23
8) 8
9) 48
10) 50
11) 14.2
12) 18
13) Yes
14) Yes
15) Yes
16) No
17) No
18) No
19) Yes
20) Yes
21) No
22) No
23) Yes
24) Yes
25) 24
26) 392
27) 10
28) 42
29) 48
30) 96
31) 96
32) 68
33) 8.4
34) 8
35) 73.5
36) 94

Similarity and ratios

1) 30
2) 2
3) 10
4) 11
5) 150 feet
6) 8.96 meters
7) 40 feet
8) 187.2 miles

Ratio and Rates Word Problems

1) 4:5
2) 252

3) The ratio for both classes is 9 to 5.
4) Walmart is a better buy.
5) 510, the rate is 30 per hour.

6) $39.20
7) 28
8) 18

Percentage Calculations

1) 1.8
2) 1.6
3) 20
4) 12
5) 27
6) 24.5
7) 4.2
8) 96
9) 43.2
10) 61

11) 182
12) 70.4
13) 88
14) 26.32
15) 25.5
16) 69.36
17) 72
18) 121
19) 50%
20) 80%

21) 5%
22) 25%
23) 12.5%
24) 25%
25) 25%
26) 3%
27) 5%
28) 14.33%
29) 30.30%
30) 54

Percent Problems

1) 50%
2) 40%
3) 160%
4) 4%
5) 1.5%
6) 6%
7) 20%
8) 8%

9) 30%
10) 25%
11) 50%
12) 50%
13) 8%
14) 125%
15) 150%
16) 30%

17) 50%
18) 320%
19) 75%
20) 14.4 ounces
21) 60
22) 75%
23) 20%

Discount, Tax and Tip

1) $486.00
2) $249.60
3) $1,008.00
4) $10,660.00
5) $412.00
6) $370,080

7) $144.00
8) $144.00
9) $56.00
10) $640.00
11) $64.40
12) $55.00

13) $117.50
14) $62.40
15) $80.00
16) $96.00
17) $43.35
18) $540.80

Percent of Change

1) 100%
2) 400%
3) 600%
4) 300%
5) 200%
6) 150%
7) 50%
8) 50%
9) 40%
10) 50%
11) 25%
12) 60%
13) 20%
14) 25%
15) 50%
16) 15%
17) 20%

Simple Interest

1) $126.00
2) $216.00
3) $475.00
4) $56.875
5) $8.00
6) $5,880.00
7) $1,536.00
8) $105.00
9) $9.80
10) $576.00
11) $14,880.00
12) $2,160.00
13) $4,050.00
14) 2 years
15) 4 years
16) 3 years

Chapter 4:
Exponents and Radicals Expressions

Topics that you will practice in this chapter:

- ✓ Multiplication Property of Exponents
- ✓ Zero and Negative Exponents
- ✓ Division Property of Exponents
- ✓ Powers of Products and Quotients
- ✓ Negative Exponents and Negative Bases
- ✓ Scientific Notation
- ✓ Square Roots
- ✓ Simplifying Radical Expressions

Mathematics is no more computation than typing is literature.
   – John Allen Paulos

Multiplication Property of Exponents

✏️ Simplify and write the answer in exponential form.

1) $2 \times 2^5 =$

2) $7^2 \times 7 =$

3) $8^3 \times 8^3 =$

4) $9^4 \times 9^3 =$

5) $4^2 \times 4^4 \times 4 =$

6) $5 \times 5^2 \times 5^3 =$

7) $9^3 \times 9^3 \times 9 \times 9 =$

8) $4x \times x =$

9) $x^5 \times x^3 =$

10) $x^6 \times x^2 =$

11) $x^2 \times x^4 \times x^5 =$

12) $7x \times 7x =$

13) $4x^2 \times 5x^3 =$

14) $12x^3 \times x =$

15) $3x^2 \times 3x^2 \times 3x^2 =$

16) $7x^5 \times 2x^3 =$

17) $x^8 \times 2x =$

18) $3x \times 3x^3 =$

19) $6x^2 \times 2x^5 =$

20) $3yx^3 \times 12x =$

21) $8x^3 \times y^5 x^2 =$

22) $4y^7 x^2 \times 3y^2 x^5 =$

23) $9yx^2 \times 4x^5 y^2 =$

24) $10x^4 \times 11x^4 y^4 =$

25) $9x^3 y^4 \times 9x^6 y^2 =$

26) $12x^4 y^4 \times 6xy^3 =$

27) $9xy^4 \times 11x^3 y^3 =$

28) $6x^2 y^4 \times 8x^3 y^6 =$

29) $8x \times y^7 x^2 \times 5y^3 =$

30) $3yx^3 \times 2y^3 x^2 \times 7xy =$

31) $8yx^5 \times 3y^4 x \times 3xy^3 =$

32) $9x^3 \times 11y^4 x^3 \times 2yx^4 =$

WWW.MathNotion.Com

Zero and Negative Exponents

✏️ **Evaluate the following expressions.**

1) $1^{-5} =$

2) $2^{-4} =$

3) $2^{-5} =$

4) $3^0 =$

5) $3^{-2} =$

6) $2^{-7} =$

7) $13^{-2} =$

8) $14^{-2} =$

9) $2^{-8} =$

10) $20^{-2} =$

11) $19^{-1} =$

12) $3^{-6} =$

13) $15^{-2} =$

14) $10^{-2} =$

15) $16^{-2} =$

16) $30^{-2} =$

17) $8^{-4} =$

18) $3^{-7} =$

19) $2^{-10} =$

20) $10^{-3} =$

21) $18^{-2} =$

22) $25^{-2} =$

23) $40^{-2} =$

24) $50^{-2} =$

25) $11^{-3} =$

26) $22^{-2} =$

27) $17^{-2} =$

28) $3^{-8} =$

29) $4^{-5} =$

30) $60^{-2} =$

31) $\left(\frac{1}{3}\right)^{-2}$

32) $\left(\frac{1}{5}\right)^{-3} =$

33) $\left(\frac{1}{8}\right)^{-2} =$

34) $\left(\frac{2}{5}\right)^{-2} =$

35) $\left(\frac{1}{15}\right)^{-2} =$

36) $\left(\frac{7}{12}\right)^{-2} =$

37) $\left(\frac{1}{20}\right)^{-2} =$

38) $\left(\frac{1}{7}\right)^{-3} =$

39) $\left(\frac{2}{3}\right)^{-5} =$

40) $\left(\frac{9}{11}\right)^{-1} =$

41) $\left(\frac{8}{9}\right)^{-2} =$

42) $\left(\frac{1}{8}\right)^{-3} =$

Division Property of Exponents

✎ **Simplify.**

1) $\dfrac{5^2}{5^6} =$

2) $\dfrac{6^9}{6^5} =$

3) $\dfrac{9^7}{9} =$

4) $\dfrac{3}{3^3} =$

5) $\dfrac{2x}{x^8} =$

6) $\dfrac{4 \times 4^5}{4^5 \times 4^2} =$

7) $\dfrac{12^6}{12^2} =$

8) $\dfrac{7 \times 7^9}{7^2 \times 7^4} =$

9) $\dfrac{4^5 \times 4^8}{4^2 \times 4^{11}} =$

10) $\dfrac{20x}{40x^4} =$

11) $\dfrac{8x^9}{9x^6} =$

12) $\dfrac{24x^3}{16x^5} =$

13) $\dfrac{25x^2}{50y^8} =$

14) $\dfrac{60xy^5}{12x^4y^2} =$

15) $\dfrac{8x^7}{12x} =$

16) $\dfrac{48x^2y^4}{16x^5} =$

17) $\dfrac{50x^6}{25x^9y^{14}} =$

18) $\dfrac{90yx^8}{15yx^9} =$

19) $\dfrac{18x^9y}{36x^{12}y^3} =$

20) $\dfrac{9x^8}{81x^8} =$

21) $\dfrac{9x^{-7}}{11x^{-3}} =$

Powers of Products and Quotients

✎ **Simplify.**

1) $(4^2)^3 =$

2) $(5^2)^2 =$

3) $(3 \times 3^2)^3 =$

4) $(3 \times 2^3)^2 =$

5) $(15^2 \times 15^2)^5 =$

6) $(7^2 \times 7^3)^4 =$

7) $(9 \times 9^2)^2 =$

8) $(4^6)^3 =$

9) $(7x^7)^3 =$

10) $(8x^4y^3)^2 =$

11) $(3x^3y^2)^4 =$

12) $(4x^2y^2)^2 =$

13) $(3x^5y^2)^3 =$

14) $(4x^3y^2)^3 =$

15) $(2x^3x)^5 =$

16) $(6x^4x^2)^2 =$

17) $(7x^{12}y^5)^2 =$

18) $(5x^7x^4)^3 =$

19) $(8x^2 \times 6x)^2 =$

20) $(9x^{14}y^3)^3 =$

21) $(5x^4y^2)^4 =$

22) $(3x^3y^7)^5 =$

23) $(8x \times 2y^3)^2 =$

24) $\left(\dfrac{8x}{x^3}\right)^3 =$

25) $\left(\dfrac{x^4y^5}{x^3y^5}\right)^7 =$

26) $\left(\dfrac{36xy}{6x^5}\right)^2 =$

27) $\left(\dfrac{x^4}{x^5y^2}\right)^3 =$

28) $\left(\dfrac{xy^2}{x^3y^8}\right)^{-3} =$

29) $\left(\dfrac{5xy^7}{x^2}\right)^3 =$

30) $\left(\dfrac{xy^5}{2xy^3}\right)^{-6} =$

Negative Exponents and Negative Bases

✏️ **Simplify.**

1) $-4^{-2} =$

2) $-7^{-1} =$

3) $-5^{-2} =$

4) $-x^{-9} =$

5) $10x^{-2} =$

6) $-7x^{-4} =$

7) $-15x^{-4} =$

8) $-15x^{-7}y^{-4} =$

9) $32x^{-9}y^{-3} =$

10) $45a^{-7}b^{-3} =$

11) $-25x^3y^{-5} =$

12) $-\dfrac{18}{x^{-9}} =$

13) $-\dfrac{13x}{a^{-8}} =$

14) $\left(-\dfrac{1}{3}\right)^{-4} =$

15) $\left(-\dfrac{3}{4}\right)^{-3} =$

16) $-\dfrac{12}{a^{-6}b^{-4}} =$

17) $-\dfrac{48x}{x^{-6}} =$

18) $-\dfrac{a^{-12}}{b^{-5}} =$

19) $-\dfrac{27}{x^{-5}} =$

20) $\dfrac{12b}{-48c^{-6}} =$

21) $\dfrac{24ab}{a^{-4}b^{-3}} =$

22) $-\dfrac{8n^{-7}}{40p^{-9}} =$

23) $\dfrac{9ab^{-6}}{-5c^{-2}} =$

24) $\left(\dfrac{2a}{3c}\right)^{-4} =$

25) $\left(-\dfrac{8x}{5yz}\right)^{-2} =$

26) $\dfrac{9ab^{-6}}{-4c^{-3}} =$

27) $\left(-\dfrac{x^3}{x^4}\right)^{-5} =$

28) $\left(-\dfrac{x^{-3}}{3x^3}\right)^{-3} =$

29) $\left(-\dfrac{x^{-6}}{x^4}\right)^{-3} =$

Scientific Notation

✍ **Write each number in scientific notation.**

1) $0.226 =$

2) $0.05 =$

3) $4.8 =$

4) $90 =$

5) $120 =$

6) $0.123 =$

7) $82 =$

8) $5,400 =$

9) $2,460 =$

10) $75,300 =$

11) $61,000,000 =$

12) $0.00009 =$

13) $468,000 =$

14) $0.00458 =$

15) $0.000087 =$

16) $31,800,000 =$

17) $950,000 =$

18) $9,000,000,000 =$

19) $0.0007 =$

20) $0.00041 =$

✍ **Write each number in standard notation.**

21) $4 \times 10^{-2} =$

22) $7 \times 10^{-4} =$

23) $4.3 \times 10^{6} =$

24) $7 \times 10^{-4} =$

25) $8.7 \times 10^{-3} =$

26) $12 \times 10^{5} =$

27) $35 \times 10^{3} =$

28) $1.89 \times 10^{5} =$

29) $13 \times 10^{-6} =$

30) $7.3 \times 10^{-4} =$

Square Roots

✎ Find the value each square root.

1) $\sqrt{64} = \underline{}$

2) $\sqrt{4} = \underline{}$

3) $\sqrt{289} = \underline{}$

4) $\sqrt{0.25} = \underline{}$

5) $\sqrt{0.01} = \underline{}$

6) $\sqrt{0.09} = \underline{}$

7) $\sqrt{1,600} = \underline{}$

8) $\sqrt{2.25} = \underline{}$

9) $\sqrt{0} = \underline{}$

10) $\sqrt{0.04} = \underline{}$

11) $\sqrt{0.36} = \underline{}$

12) $\sqrt{0.81} = \underline{}$

13) $\sqrt{0.49} = \underline{}$

14) $\sqrt{1.21} = \underline{}$

15) $\sqrt{1.69} = \underline{}$

16) $\sqrt{0.16} = \underline{}$

17) $\sqrt{529} = \underline{}$

18) $\sqrt{625} = \underline{}$

19) $\sqrt{0.81} = \underline{}$

20) $\sqrt{20} = \underline{}$

21) $\sqrt{50} = \underline{}$

22) $\sqrt{676} = \underline{}$

23) $\sqrt{270} = \underline{}$

24) $\sqrt{32} = \underline{}$

✎ Evaluate.

25) $\sqrt{4} \times \sqrt{16} = \underline{}$

26) $\sqrt{49} \times \sqrt{64} = \underline{}$

27) $\sqrt{2} \times \sqrt{8} = \underline{}$

28) $\sqrt{17} \times \sqrt{17} = \underline{}$

29) $\sqrt{13} \times \sqrt{13} = \underline{}$

30) $\sqrt{15} \times \sqrt{15} = \underline{}$

31) $\sqrt{19} + \sqrt{19} = \underline{}$

32) $\sqrt{1} + \sqrt{1} = \underline{}$

33) $8\sqrt{7} - 2\sqrt{7} = \underline{}$

34) $7\sqrt{10} \times 6\sqrt{10} = \underline{}$

35) $9\sqrt{5} \times 2\sqrt{5} = \underline{}$

36) $8\sqrt{3} - \sqrt{12} = \underline{}$

Simplifying Radical Expressions

✎ **Simplify.**

1) $\sqrt{13y^2} =$

2) $\sqrt{60x^3} =$

3) $\sqrt[3]{27a} =$

4) $\sqrt{81x^2} =$

5) $\sqrt{150a} =$

6) $\sqrt[3]{135w^3} =$

7) $\sqrt{200x} =$

8) $\sqrt{192v} =$

9) $\sqrt[3]{64x} =$

10) $\sqrt{84x^3} =$

11) $\sqrt{121x^2} =$

12) $\sqrt[3]{48a} =$

13) $\sqrt{480} =$

14) $\sqrt{1,575p^2} =$

15) $\sqrt{108m^6} =$

16) $\sqrt{198x^3y^2} =$

17) $\sqrt{169x^2y^3} =$

18) $\sqrt{25a^6} =$

19) $\sqrt{50x^2y^3} =$

20) $\sqrt[3]{512y^3} =$

21) $2\sqrt{144x^2} =$

22) $3\sqrt{400x^2} =$

23) $\sqrt[3]{189xy^4} =$

24) $\sqrt[3]{1,331x^3y^5} =$

25) $3\sqrt{150a} =$

26) $\sqrt[3]{729y} =$

27) $3\sqrt{18xyr^3} =$

28) $6\sqrt{225x^2yz^6} =$

29) $3\sqrt[3]{125x^3y^2} =$

30) $7\sqrt{12a^2bc^4} =$

31) $4\sqrt[3]{1,000x^9y^{15}} =$

Answers of Worksheets – Chapter 4

Multiplication Property of Exponents

1) 2^6
2) 7^3
3) 8^6
4) 9^7
5) 4^7
6) 5^6
7) 9^8
8) $4x^2$
9) x^8
10) x^8
11) x^{11}
12) $49x^2$
13) $20x^5$
14) $12x^4$
15) $27x^6$
16) $14x^8$
17) $2x^8$
18) $9x^4$
19) $12x^7$
20) $36x^4y$
21) $8x^5y^5$
22) $12x^7y^9$
23) $36x^7y^3$
24) $110x^8y^4$
25) $81x^9y^6$
26) $72x^5y^7$
27) $99x^4y^7$
28) $48x^5y^{10}$
29) $40x^3y^{10}$
30) $42x^6y^5$
31) $72x^7y^8$
32) $198x^{10}y^5$

Zero and Negative Exponents

1) 1
2) $\frac{1}{16}$
3) $\frac{1}{32}$
4) 1
5) $\frac{1}{9}$
6) $\frac{1}{128}$
7) $\frac{1}{169}$
8) $\frac{1}{196}$
9) $\frac{1}{256}$
10) $\frac{1}{400}$
11) $\frac{1}{19}$
12) $\frac{1}{729}$
13) $\frac{1}{225}$
14) $\frac{1}{100}$
15) $\frac{1}{256}$
16) $\frac{1}{900}$
17) $\frac{1}{4,096}$
18) $\frac{1}{2,187}$
19) $\frac{1}{1,024}$
20) $\frac{1}{1,000}$
21) $\frac{1}{324}$
22) $\frac{1}{625}$
23) $\frac{1}{1,600}$
24) $\frac{1}{2,500}$
25) $\frac{1}{1,331}$
26) $\frac{1}{484}$
27) $\frac{1}{289}$
28) $\frac{1}{6,561}$
29) $\frac{1}{1,024}$
30) $\frac{1}{3,600}$
31)
32)
33)
34) 6.25
35) 225
36) $\frac{144}{49}$
37) 400
38) 343
39) $\frac{243}{32}$
40) $\frac{11}{9}$
41) $\frac{81}{64}$
42) 512

Division Property of Exponents

1) $\frac{1}{5^4}$
2) 6^4
3) 9^6
4) $\frac{1}{3^2}$
5) $\frac{2}{x^7}$
6) $\frac{1}{4}$
7) 12^4
8) 7^4
9) 1
10) $\frac{1}{2x^3}$
11) $\frac{8x^3}{9}$
12) $\frac{3}{2x^2}$
13) $\frac{x^2}{2y^8}$
14) $\frac{5y^3}{x^3}$
15) $\frac{2x^6}{3}$
16) $\frac{3y^4}{x^3}$
17) $\frac{2}{x^3y^{14}}$

18) $\frac{6}{x}$ 19) $\frac{1}{2x^3y^2}$ 20) $\frac{1}{9}$ 21) $\frac{9}{11x^4}$

Powers of Products and Quotients

1) 4^6
2) 5^4
3) 3^9
4) 24^2
5) 15^{20}
6) 7^{20}
7) 9^6
8) 4^{18}
9) $343x^{21}$
10) $64x^8y^6$
11) $81x^{12}y^8$
12) $16x^4y^4$
13) $27x^{15}y^6$
14) $64x^9y^6$
15) $32x^{20}$
16) $36x^{12}$
17) $49x^{24}y^{10}$
18) $125x^{33}$
19) $2{,}304x^6$
20) $729x^{42}y^9$
21) $625x^{16}y^9$
22) $243x^{15}y^8$
23) $256x^2y^6$
24) $\frac{512}{x^6}$
25) x^7
26) $\frac{36y^2}{x^8}$
27) $\frac{1}{x^3y^6}$
28) x^6y^{18}
29) $\frac{125y^{21}}{x^3}$
30) $\frac{64}{y^{12}}$

Negative Exponents and Negative Bases

1) $-\frac{1}{16}$
2) $-\frac{1}{7}$
3) $-\frac{1}{25}$
4) $-\frac{1}{x^9}$
5) $\frac{10}{x^2}$
6) $-\frac{7}{x^4}$
7) $-\frac{15}{x^4}$
8) $-\frac{15}{x^7y^4}$
9) $\frac{32}{x^9y^3}$
10) $\frac{45}{a^7b^3}$
11) $-\frac{25x^3}{y^5}$
12) $-18x^9$
13) $-13xa^8$
14) 81
15) $-\frac{64}{27}$
16) $-12a^6b^4$
17) $-48x^7$
18) $-\frac{b^5}{a^{12}}$
19) $-27x^5$
20) $-\frac{bc^6}{4}$
21) $24a^5b^4$
22) $-\frac{p^9}{5n^7}$
23) $-\frac{9ac^2}{5b^6}$
24) $\frac{81c^4}{16a^4}$
25) $\frac{25y^2z^2}{64x^2}$
26) $-\frac{9ac^3}{4b^6}$
27) $-x^5$
28) $-27x^{18}$
29) $-x^{30}$

Writing Scientific Notation

1) 2.26×10^{-1}
2) 5×10^{-2}
3) 4.8×10^0
4) 9×10^1
5) 1.2×10^2
6) 1.23×10^{-1}
7) 8.2×10^1
8) 5.4×10^3
9) 2.46×10^3

10) 7.53×10^4
11) 61×10^6
12) 9×10^{-5}
13) 4.68×10^5
14) 4.58×10^{-3}
15) 8.7×10^{-5}
16) 3.18×10^7

17) 9.5×10^5
18) 9×10^9
19) 7×10^{-4}
20) 4.1×10^{-4}
21) 0.04
22) 0.0007
23) 4,300,000

24) 0.0007
25) 0.0087
26) 1,200,000
27) 35,000
28) 189,000
29) 0.000013
30) 0.00073

Square Roots

1) 8
2) 2
3) 17
4) 0.5
5) 0.1
6) 0.3
7) 40
8) 1.5
9) 0
10) 0.2

11) 0.6
12) 0.9
13) 0.7
14) 1.1
15) 1.3
16) 0.4
17) 23
18) 25
19) 0.9
20) $2\sqrt{5}$

21) $5\sqrt{2}$
22) 26
23) $3\sqrt{30}$
24) $4\sqrt{2}$
25) 8
26) 56
27) 4
28) 17
29) 13

30) 15
31) $2\sqrt{19}$
32) 2
33) $6\sqrt{7}$
34) 420
35) 90
36) $6\sqrt{3}$

Simplifying radical expressions

1) $y\sqrt{13}$
2) $2x\sqrt{15x}$
3) $3\sqrt[3]{a}$
4) $9x$
5) $5\sqrt{6a}$
6) $3w\sqrt[3]{5}$
7) $10\sqrt{2x}$
8) $8\sqrt{3v}$
9) $4\sqrt[3]{x}$
10) $2x\sqrt{21x}$
11) $11x$

12) $2\sqrt[3]{6a}$
13) $4\sqrt{30}$
14) $15p\sqrt{7}$
15) $6m^3\sqrt{3}$
16) $3x.y\sqrt{22x}$
17) $13xy\sqrt{y}$
18) $5a^3$
19) $5xy\sqrt{2y}$
20) $8y$
21) $24x$
22) $60x$

23) $3y\sqrt[3]{7xy}$
24) $11xy\sqrt[3]{y^2}$
25) $15\sqrt{6a}$
26) $9\sqrt[3]{y}$
27) $9r\sqrt{2xyr}$
28) $90xz^3\sqrt{y}$
29) $15x\sqrt[3]{y^2}$
30) $14ac^2\sqrt{b}$
31) $40x^3y^{15}$

Chapter 5:
Algebraic Expressions

Topics that you will practice in this chapter:

- ✓ Simplifying Variable Expressions
- ✓ Simplifying Polynomial Expressions
- ✓ Translate Phrases into an Algebraic Statement
- ✓ The Distributive Property
- ✓ Evaluating One Variable Expressions
- ✓ Evaluating Two Variables Expressions
- ✓ Combining like Terms

Mathematics is, as it were, a sensuous logic, and relates to philosophy as do the arts, music, and plastic art to poetry. — *K. Shegel*

Simplifying Variable Expressions

✏️ **Simplify each expression.**

1) $3(x + 8) =$

2) $(-4)(7x - 3) =$

3) $11x + 8 - 7x =$

4) $-6 - 2x^2 - 9x^2 =$

5) $8 + 17x^2 + 6 =$

6) $9x^2 + 13x + 19x^2 =$

7) $7x^2 - 15x^2 + 3x =$

8) $8x^2 - 11x - 3x =$

9) $3x + 9(1 - 4x) =$

10) $14x + 2(20x - 4) =$

11) $6(-3x - 7) - 22 =$

12) $7x^2 + (-12x) =$

13) $x - 8 + 15 - 7x =$

14) $3 - 6x + 12 - 3x =$

15) $20x - 14 + 27 + 12x =$

16) $(-7)(6x - 5) + 14x =$

17) $11x - 4(3 - 7x) =$

18) $22x + 3(5x + 2) + 11 =$

19) $4(-3x + 8) + 10x =$

20) $16x - 3x(2x + 7) =$

21) $9x + 12x(2 - 4x) =$

22) $5x(-4x + 11) + 18x =$

23) $20x + 24x + 3x^2 =$

24) $7x(x - 7) - 28 =$

25) $7x - 12 + 5x + 3x^2 =$

26) $4x^2 - 9x - 12x =$

27) $8x - 22x^2 - 21x^2 - 11 =$

28) $9 + 3x^2 - 8x^2 - 27x =$

29) $14x + 2x^2 + 4x + 28 =$

30) $7x^2 + 45x + 12x^2 =$

31) $25 + 15x^2 + 9x - 3x^2 =$

32) $17x - 32x - 2x^2 + 30 =$

Simplifying Polynomial Expressions

✏️ **Simplify each polynomial.**

1) $(5x^4 + 2x^2) - (11x + 6x^2) =$ _____

2) $(x^7 + 6x^4) - (8x^4 + 4x^2) =$ _____

3) $(24x^5 + 8x^3) - (x^3 - 12x^5) =$ _____

4) $14x - 9x^5 - 4(7x^5 + 7x^3) =$ _____

5) $(7x^4 - 5) + 2(4x^2 - 8x^4) =$ _____

6) $(9x^5 - 3x) - 3(8x^5 - 3x^4) =$ _____

7) $4(2x - 4x^4) - 5(3x^4 + x^2) =$ _____

8) $(4x^2 - 2x) - (5x^3 + 9x^2) =$ _____

9) $8x^4 - (9x^6 + 2x) + 2x^2 =$ _____

10) $x^5 - 3(x^3 + 2x) + 9x =$ _____

11) $(4x^2 - 2x^5) - (4x^5 - 2x^2) =$ _____

12) $8x^3 - 8x^5 + 17x^4 - 12x^5 =$ _____

13) $4x^3 - 9x^7 + 18x^7 - 24x^6 =$ _____

14) $4x^5 + 13x^3 - 17x^5 + 24x =$ _____

15) $7x^6 - 9x^7 + 5x^6 - 12x^3 =$ _____

16) $4x^4 + 19x - 3x^3 - 21x^4 =$ _____

Translate Phrases into an Algebraic Statement

✎ Write an algebraic expression for each phrase.

1) 13 multiplied by x. _____

2) Subtract 15 from y. _____

3) 22 divided by x. _____

4) 27 decreased by y. _____

5) Add y to 31. _____

6) The square of 7. _____

7) x raised to the seventh power. _____

8) The sum of five and a number. _____

9) The difference between forty–nine and y. _____

10) The quotient of eight and a number. _____

11) The quotient of the square of x and 34. _____

12) The difference between x and 14 is 41. _____

13) 7 times b reduced by the square of a. _____

14) Subtract the product of a and b from 51. _____

The Distributive Property

✏️ **Use the distributive property to simply each expression.**

1) $4(2 + 5x) =$

2) $5(2 + 4x) =$

3) $6(5x - 5) =$

4) $(6x - 3)(-7) =$

5) $(-4)(x + 8) =$

6) $(4 + 4x)6 =$

7) $(-5)(8 - 7x) =$

8) $-(-3 - 12x) =$

9) $(-8x + 3)(-5) =$

10) $(-5)(x - 11) =$

11) $-(8 - 2x) =$

12) $3(7 + 4x) =$

13) $4(8 + 3x) =$

14) $(-8x + 2)5 =$

15) $(4 - 7x)(-9) =$

16) $(-12)(3x + 5) =$

17) $(9 - 3x)5 =$

18) $4(4 + 7x) =$

19) $12(3x - 6) =$

20) $(-7x + 5)4 =$

21) $(4 - 9x)(-2) =$

22) $(-15)(2x - 3) =$

23) $(14 - 3x)3 =$

24) $(-5)(10x - 4) =$

25) $(5 - 7x)(-12) =$

26) $(-8)(2x + 9) =$

27) $(-5 + 8x)(-7) =$

28) $(-6)(2 - 15x) =$

29) $13(4x - 6) =$

30) $(-15x + 13)(-4) =$

31) $(-9)(3x - 2) + 2(x + 5) =$

32) $(-9)(2x + 2) - (7 + 4x) =$

Evaluating One Variable Expressions

✎ **Evaluate each expression using the value given.**

1) $8 - x, x = 5$

2) $x - 10, x = 6$

3) $3x - 6, x = 5$

4) $x - 15, x = -2$

5) $12 - x, x = 4$

6) $x + 7, x = 1$

7) $2x + 9, x = 7$

8) $x + (-4), x = -7$

9) $2x + 9, x = 4$

10) $3x + 10, x = -2$

11) $18 + 2x - 4, x = -1$

12) $18 - 6x, x = 2$

13) $8x - 2, x = 4$

14) $2x - 17, x = 8$

15) $13x - 12, x = 3$

16) $8 - 5x, x = -2$

17) $3(5x + 4), x = 5$

18) $4(-2x - 7), x = 3$

19) $7x - 5x + 12, x = 2$

20) $(8x + 4) \div 2, x = 6$

21) $(x + 15) \div 4, x = 9$

22) $6x - 10 + 3x, x = -5$

23) $(7 - 4x)(-3), x = -3$

24) $12x^2 + 5x - 4, x = 2$

25) $x^2 - 15x, x = -4$

26) $3x(3 - 6x), x = 2$

27) $13x + 8 - 6x^2, x = -3$

28) $(-2)(15x - 11 + 4x), x = 4$

29) $(-6) + \frac{x}{6} + x, x = 18$

30) $(-9) + \frac{x}{4}, x = 32$

31) $\left(-\frac{45}{x}\right) - 5 + 2x, x = 9$

32) $\left(-\frac{36}{x}\right) - 9 + 3x, x = 3$

Evaluating Two Variables Expressions

✎ **Evaluate each expression using the values given.**

1) $5x - y$,
   $x = 4, y = 3$

2) $3x + 2y$,
   $x = -2, y = 2$

3) $-6a + 5b$,
   $a = 3, b = 1$

4) $3x + 7 - y$,
   $x = 8, y = 4$

5) $5z + 12 - 3k$,
   $z = 5, k = 2$

6) $6(-x - 3y)$,
   $x = 5, y = 4$

7) $7a + 4b$,
   $a = 3, b = 5$

8) $8x \div 4y$,
   $x = 6, y = 4$

9) $2x + 18 + 4y$,
   $x = -3, y = 3$

10) $5a - (18 - 2b)$,
    $a = 5, b = 8$

11) $6z + 12 + 3k$,
    $z = -3, k = 3$

12) $2xy + 6 + 7x$,
    $x = 5, y = 3$

13) $6x + 2y - 9 + 3$,
    $x = 3, y = 2$

14) $\left(-\frac{21}{x}\right) + 6 + 3y$,
    $x = 7, y = 4$

15) $(-4)(-3a - b)$,
    $a = 2, b = 6$

16) $18 + 4x + 9 - 5y$,
    $x = 6, y = 4$

17) $7x + 5 - 6y + 11$,
    $x = 9, y = 3$

18) $9 + 4(-5x - 3y)$,
    $x = 4, y = 5$

19) $3x + 15 + 6y$,
    $x = 2, y = 4$

20) $7a - (4a - 2b) + 8$,
    $a = 3, b = 1$

Combining like Terms

✎ **Simplify each expression.**

1) $7x + 2x + 8 =$

2) $3(6x - 2) =$

3) $10x - 12x + 8 =$

4) $20x - 32x + 14 =$

5) $16x - 6x - 12 =$

6) $18x - 21 + 4x =$

7) $15 - (3x + 9) =$

8) $-14x + 7 - 11x =$

9) $5x - 10 - 3x + 1 =$

10) $24x + 7x - 22 =$

11) $14x + 8x - 2 =$

12) $(-4x + 2)8 =$

13) $34 + 6x + 8x - 4 =$

14) $3(x - 8x) - 5 =$

15) $4(2x + 7) + 5x =$

16) $x - 27 - 9x =$

17) $3(5 + 4x) - 8x =$

18) $41x + 24 + 3x =$

19) $(-8x) + 30 + 15x =$

20) $(-4x) - 12 + 19x =$

21) $5(2x + 6) + 9x =$

22) $3(6 - 7x) - 11x =$

23) $-8x - (16 - 14x) =$

24) $(-9) - (6)(5x + 9) =$

25) $(-4)(6x - 5) - 12x =$

26) $-34x + 14 + 9x - 21x =$

27) $5(-13x + 6) - 24x =$

28) $-7x - 20 + 15x =$

29) $42x - 31x + 15 - 12x =$

30) $4(8x + 5x) - 17 =$

31) $54 - 22x - 28 - 19x =$

32) $-9(-7x - 11x) + 58x =$

Answers of Worksheets – Chapter 5

Simplifying Variable Expressions

1) $3x + 24$
2) $-28x + 12$
3) $4x + 8$
4) $-11x^2 - 6$
5) $17x^2 + 14$
6) $28x^2 + 13x$
7) $-8x^2 + 3x$
8) $8x^2 - 14x$
9) $-33x + 9$
10) $54x - 8$
11) $-18x - 64$
12) $7x^2 - 12x$
13) $-6x + 7$
14) $-9x + 15$
15) $32x + 13$
16) $-28x + 35$
17) $39x - 12$
18) $37x + 17$
19) $-2x + 32$
20) $-6x^2 - 5x$
21) $-48x^2 + 33x$
22) $-20x^2 + 73x$
23) $3x^2 + 44x$
24) $7x^2 - 49x - 28$
25) $3x^2 + 12x - 12$
26) $4x^2 - 21x$
27) $-43x^2 + 8x - 11$
28) $-5x^2 - 27x + 9$
29) $2x^2 + 18x + 28$
30) $19x^2 + 45x$
31) $12x^2 + 9x + 25$
32) $-2x^2 - 15x + 30$

Simplifying Polynomial Expressions

1) $5x^4 - 4x^2 - 11x$
2) $x^7 - 2x^4 - 4x^2$
3) $36x^5 + 7x^3$
4) $-37x^5 - 28x^2 + 14x$
5) $-9x^4 + 8x^2 - 5$
6) $-15x^5 + 9x^4 - 3x$
7) $-31x^3 - 5x^2 + 8x$
8) $-5x^3 - 5x^2 - 2x$
9) $-9x^6 + 8x^4 + 2x^2 - 2x$
10) $x^5 - 3x^3 + 3x$
11) $-6x^5 + 6x^2$
12) $-20x^5 + 17x^4 + 8x^3$
13) $9x^7 - 24x^6 + 4x^3$
14) $-13x^5 + 13x^3 + 24x$
15) $-9x^7 + 12x^6 - 12x^3$
16) $-17x^4 - 3x^3 + 19x$

Translate Phrases into an Algebraic Statement

1) $13x$
2) $y - 15$
3) $\frac{22}{x}$
4) $27 - y$
5) $y + 31$
6) 7^2
7) x^7
8) $5 + x$
9) $49 - y$
10) $\frac{8}{x}$
11) $\frac{x^2}{34}$
12) $x - 14 = 41$
13) $7b - a^2$
14) $51 - ab$

The Distributive Property

1) $20x + 8$
2) $20x + 10$
3) $30x - 30$
4) $-42x + 21$
5) $-4x - 32$
6) $24x + 24$
7) $35x - 40$
8) $12x + 3$

9) $40x - 15$
10) $-5x + 55$
11) $2x - 8$
12) $12x + 21$
13) $12x + 32$
14) $-40x + 10$

15) $63x - 36$
16) $-36x - 60$
17) $-15x + 45$
18) $28x + 16$
19) $36x - 72$
20) $-28x + 20$

21) $18x - 8$
22) $-30x + 45$
23) $-9x + 42$
24) $-50x + 20$
25) $84x - 60$
26) $-16x - 72$

27) $56x - 35$
28) $90x - 12$
29) $52x - 78$
30) $60x - 52$
31) $-25x + 28$
32) $-22x - 25$

Evaluating One Variables

1) 3
2) -4
3) 9
4) -17
5) 8
6) 8
7) 23
8) -11

9) 17
10) 4
11) 12
12) 6
13) 30
14) -1
15) 27
16) 18

17) 87
18) -52
19) 16
20) 26
21) 6
22) -55
23) -57
24) 54

25) 76
26) -54
27) -85
28) -130
29) 15
30) -1
31) 8
32) -12

Evaluating Two Variables

1) 17
2) -2
3) -13
4) 27
5) 31

6) -102
7) 41
8) 3
9) 24
10) 23

11) 3
12) 71
13) 16
14) 15
15) 48

16) 31
17) 61
18) -131
19) 45
20) 19

Combining like Terms

1) $9x + 8$
2) $18x - 6$
3) $-2x + 8$
4) $-12x + 14$
5) $10x - 12$
6) $22x - 22$
7) $-3x + 6$
8) $-25x + 7$

9) $2x - 9$
10) $31x - 22$
11) $22x - 2$
12) $-32x + 16$
13) $14x + 30$
14) $-21x - 5$
15) $13x + 28$
16) $-8x - 27$

17) $4x + 15$
18) $44x + 24$
19) $7x + 30$
20) $15x - 12$
21) $19x + 30$
22) $-32x + 18$
23) $6x - 16$
24) $-30x - 63$

25) $-36x + 20$
26) $-46x + 14$
27) $-89x + 30$
28) $8x - 20$
29) $-x + 15$
30) $52x - 17$
31) $-41x + 26$
32) $220x$

Chapter 6:
Equations and Inequalities

Topics that you will practice in this chapter:

- ✓ One–Step Equations
- ✓ Multi–Step Equations
- ✓ Graphing Single–Variable Inequalities
- ✓ One–Step Inequalities
- ✓ Multi-Step Inequalities
- ✓ Systems of Equations
- ✓ Systems of Equations Word Problems

"Life is a math equation. In order to gain the most, you have to know how to convert negatives into positives." – Anonymous

One-Step Equations

✏️ **Find the answer for each equation.**

1) $3x = 90, x =$ _____

2) $5x = 35, x =$ _____

3) $9x = 36, x =$ _____

4) $25x = 150, x =$ _____

5) $x + 18 = 23, x =$ _____

6) $x - 3 = 8, x =$ _____

7) $x - 7 = 4, x =$ _____

8) $x + 22 = 30, x =$ _____

9) $x - 11 = 6, x =$ _____

10) $24 = 28 + x, x =$ _____

11) $x - 5 = 7, x =$ _____

12) $9 - x = -7, x =$ _____

13) $43 = -8 + x, x =$ _____

14) $x - 23 = -38, x =$ _____

15) $x + 45 = -27, x =$ _____

16) $42 = 56 - x, x =$ _____

17) $-18 + x = -32, x =$ _____

18) $x - 13 = 7, x =$ _____

19) $35 = x - 10, x =$ _____

20) $x - 8 = -21, x =$ _____

21) $x - 54 = -20, x =$ _____

22) $x - 42 = -47, x =$ _____

23) $x - 8 = 29, x =$ _____

24) $-93 = x - 51, x =$ _____

25) $x + 15 = 37, x =$ _____

26) $108 = 12x, x =$ _____

27) $x - 33 = 27, x =$ _____

28) $x - 12 = 23, x =$ _____

29) $72 - x = 18, x =$ _____

30) $x + 34 = 58, x =$ _____

31) $21 - x = -9, x =$ _____

32) $x - 59 = -80, x =$ _____

Multi-Step Equations

✏ **Find the answer for each equation.**

1) $3x + 1 = 7$

2) $-x + 10 = 9$

3) $5x - 13 = 7$

4) $-(4 - x) = 5$

5) $3x - 8 = 16$

6) $15x - 13 = 17$

7) $3x - 28 = 2$

8) $9x + 21 = 39$

9) $14x + 17 = 45$

10) $-14(8 + x) = 70$

11) $8(10 + x) = 32$

12) $16 = -(x - 8)$

13) $5(7 - 3x) = 50$

14) $-19 = -(3x + 7)$

15) $30(3 + x) = 60$

16) $9(x - 12) = 54$

17) $-24 = 3x + 5x$

18) $5x + 28 = -2x - 7$

19) $9(5 + 4x) = -99$

20) $18 - x = -12 - 6x$

21) $4 - 4x = 28 - 2x$

22) $15 + 12x = -15 + 8x$

23) $54 = (-3x) - 8 + 8$

24) $12 = 7x - 18 + 5x$

25) $-18 = -9x - 42 + 5x$

26) $11x - 6 = -33 + 8x$

27) $8x - 42 = 3x + 3$

28) $-15 - 8x = 4(5 - x)$

29) $x - 9 = -5(9 - 2x)$

30) $14x - 65 = -x - 110$

31) $3x - 129 = -3(11 + 7x)$

32) $-7x - 20 = 2x + 43$

Graphing Single-Variable Inequalities

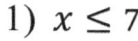 Draw a graph for each inequality.

1) $x \leq 7$

2) $x \leq -1.5$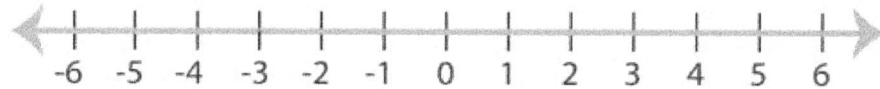

3) $x < -4$

4) $x > 2.5$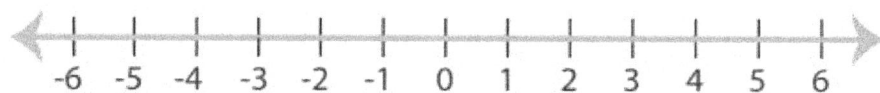

5) $x > 1.3$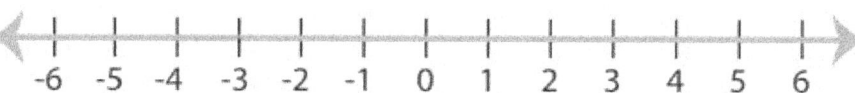

6) $x < 4$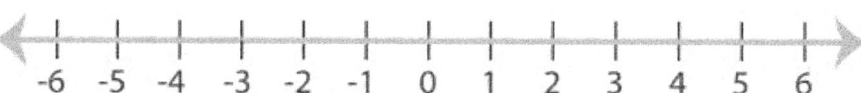

7) $x < 2.4$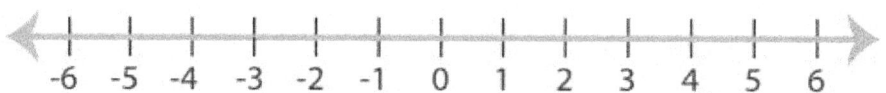

8) $x > -\frac{18}{10}$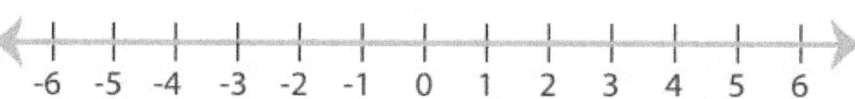

One–Step Inequalities

✏ **Find the answer for each inequality and graph it.**

1) $x + 3 > -5$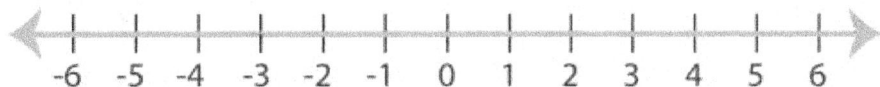

2) $x - 4 < 1$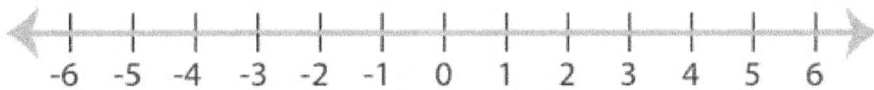

3) $7x < 42$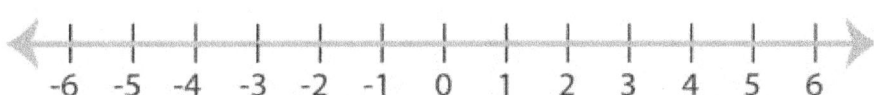

4) $13 + x > 12$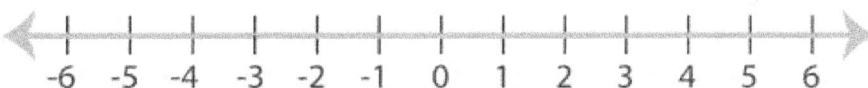

5) $x + 20 < 13$

6) $14x \le 42$

7) $11x \le -44$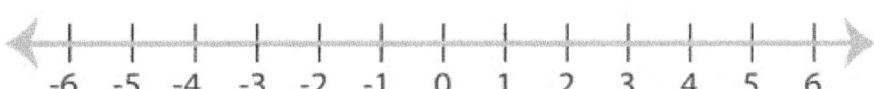

8) $x + 26 > 35$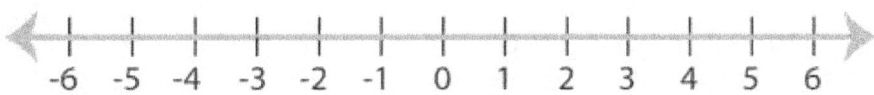

Multi-Step Inequalities

✎ **Calculate each inequality.**

1) $x - 8 \leq 12$

2) $9 - 3x \leq 18$

3) $4x - 7 \leq 9$

4) $8x - 9 \geq 15$

5) $x - 19 \geq 24$

6) $5x - 15 \leq 40$

7) $7x - 4 \leq 24$

8) $-18 + 8x \leq 22$

9) $9(x - 8) \leq 27$

10) $4x - 8 \leq 16$

11) $11x - 42 < 22$

12) $10x - 18 < 52$

13) $17 - 9x \geq -46$

14) $32 + 2x < 68$

15) $8 + 8x \geq 80$

16) $11 + 6x < 65$

17) $9x - 13 < 23$

18) $8(12 - 4x) \geq -68$

19) $-(2 + 5x) < 42$

20) $14 - 9x \geq -31$

21) $-5(x - 3) > 65$

22) $\dfrac{2x + 8}{3} \leq 12$

23) $\dfrac{8x + 16}{4} \leq 24$

24) $\dfrac{2x - 22}{9} > 8$

25) $7 + \dfrac{x}{4} < 21$

26) $\dfrac{32x}{16} - 4 < 6$

27) $\dfrac{12x + 36}{22} > 3$

28) $42 + \dfrac{x}{3} < 15$

Systems of Equations

✎ **Calculate each system of equations.**

1) $-6x + 7y = 8$ $x = $ ___
   $x + 4y = 9$ $y = $ ___

2) $-4x + 12y = 12$ $x = $ ___
   $14x - 16y = 10$ $y = $ ___

3) $y = -9$ $x = $ ___
   $2x - 5y = 12$ $y = $ ___

4) $4y = -4x + 20$ $x = $ ___
   $8x - 2y = -12$ $y = $ ___

5) $10x - 9y = -13$ $x = $ ___
   $-5x + 3y = 11$ $y = $ ___

6) $-6x - 8y = 10$ $x = $ ___
   $4x - 8y = 20$ $y = $ ___

7) $5x - 14y = -23$ $x = $ ___
   $-6x + 7y = 8$ $y = $ ___

8) $-4x + 3y = 3$ $x = $ ___
   $-x + 2y = 5$ $y = $ ___

9) $-4x + 5y = 15$ $x = $ ___
   $-3x + 4y = -10$ $y = $ ___

10) $-6x - 6y = -21$ $x = $ ___
    $-6x + 6y = -66$ $y = $ ___

11) $12x - 21y = 6$ $x = $ ___
    $-6x - 3y = -12$ $y = $ ___

12) $-4x - 4y = -14$ $x = $ ___
    $4x - 4y = 44$ $y = $ ___

13) $4x + 5y = 3$ $x = $ ___
    $3x - y = 6$ $y = $ ___

14) $3x - 2y = 2$ $x = $ ___
    $10x - 10y = 20$ $y = $ ___

15) $5x + 8y = 14$ $x = $ ___
    $-3x - 2y = -3$ $y = $ ___

16) $8x + 5y = 4$ $x = $ ___
    $-3x - 4y = 15$ $y = $ ___

Systems of Equations Word Problems

✍ **Find the answer for each word problem.**

1) Tickets to a movie cost $6 for adults and $4 for students. A group of friends purchased 9 tickets for $50.00. How many adults ticket did they buy? _____

2) At a store, Eva bought two shirts and five hats for $77.00. Nicole bought three same shirts and four same hats for $84.00. What is the price of each shirt? _____

3) A farmhouse shelters 10 animals, some are pigs, and some are ducks. Altogether there are 36 legs. How many pigs are there? _____

4) A class of 85 students went on a field trip. They took 24 vehicles, some cars and some buses. If each car holds 3 students and each bus hold 16 students, how many buses did they take? _____

5) A theater is selling tickets for a performance. Mr. Smith purchased 8 senior tickets and 10 child tickets for $248 for his friends and family. Mr. Jackson purchased 4 senior tickets and 6 child tickets for $132. What is the price of a senior ticket? $_____

6) The difference of two numbers is 15. Their sum is 33. What is the bigger number? $_____

7) The sum of the digits of a certain two-digit number is 7. Reversing its digits increase the number by 9. What is the number? _____

8) The difference of two numbers is 11. Their sum is 25. What are the numbers? _____

9) The length of a rectangle is 5 meters greater than 2 times the width. The perimeter of rectangle is 28 meters. What is the length of the rectangle? _____

10) Jim has 23 nickels and dimes totaling $2.40. How many nickels does he have? _____

Answers of Worksheets – Chapter 6

One–Step Equations

1) 30	9) 17	17) −14	25) 22
2) 7	10) −4	18) 20	26) 9
3) 4	11) 12	19) 45	27) 60
4) 6	12) 16	20) −13	28) 35
5) 5	13) 51	21) 34	29) 54
6) 11	14) −15	22) −5	30) 24
7) 11	15) −72	23) 37	31) 30
8) 8	16) 14	24) −42	32) −21

Multi–Step Equations

1) 2	9) 2	17) −3	25) −6
2) 1	10) −13	18) −5	26) −9
3) 4	11) −6	19) −4	27) 9
4) 9	12) −8	20) −6	28) −8.75
5) 8	13) −1	21) −12	29) 4
6) 2	14) 4	22) −7.5	30) −3
7) 10	15) −1	23) −18	31) 4
8) 2	16) 18	24) 2.5	32) −7

Graphing Single–Variable Inequalities

1)

2)

3)

4)

5)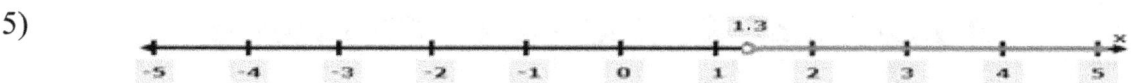

6)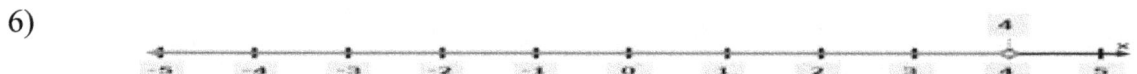

7)

8)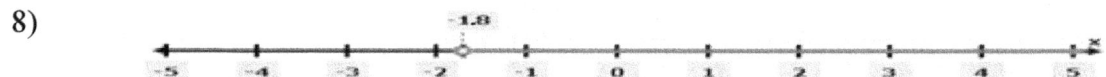

One–Step Inequalities

1)

2)

3)

4)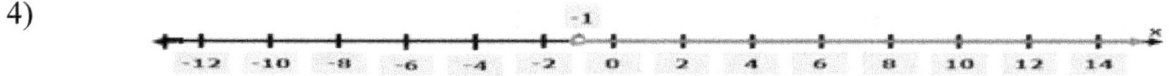

5)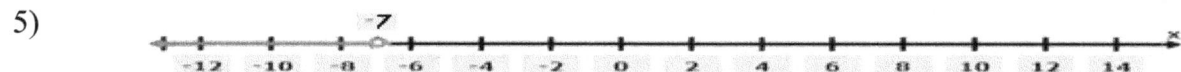

6)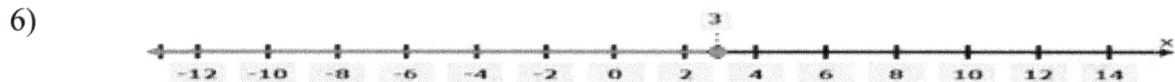

7)

8)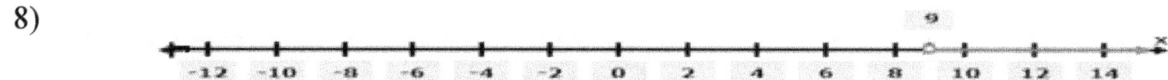

Multi-Step Inequalities

1) $x \leq 20$ 5) $x \geq 43$ 9) $x \leq 11$ 13) $x \leq 7$

2) $x \geq -3$ 6) $x \leq 11$ 10) $x \leq 6$ 14) $x < 18$

3) $x \leq 4$ 7) $x \leq 4$ 11) $x < 64/11$ 15) $x \geq 9$

4) $x \geq 3$ 8) $x \leq 5$ 12) $x < 7$ 16) $x < 9$

17) $x < 4$
18) $x \leq 41/8$
19) $x > -44/5$

20) $x \leq 5$
21) $x < -10$
22) $x \leq 14$

23) $x \leq 10$
24) $x > 47$
25) $x < 56$

26) $x < 9/4$
27) $x > 2.5$
28) $x < -81$

Systems of Equations

1) $x = 1, y = 2$
2) $x = 3, y = 2$
3) $x = -\frac{33}{2}$
4) $x = -\frac{1}{5}, y = \frac{26}{5}$
5) $x = -4, y = -3$
6) $x = 1, y = -2$

7) $x = 1, y = 2$
8) $x = \frac{9}{5}, y = \frac{17}{5}$
9) $x = -110, y = -85$
10) $x = -\frac{15}{4}, y = \frac{29}{4}$
11) $x = \frac{5}{3}, y = \frac{2}{3}$
12) $x = -\frac{15}{4}, y = \frac{29}{4}$

13) $x = \frac{33}{19}, y = -\frac{15}{19}$
14) $x = -2, y = -4$
15) $x = -\frac{2}{7}, y = \frac{27}{14}$
16) $x = \frac{91}{17}, y = -\frac{132}{17}$

Systems of Equations Word Problems

1) 7
2) $16
3) 8
4) 1

5) $21
6) 24
7) 43
8) 18, 7

9) 11 meters
10) 18

Chapter 7:

Polynomials

Topics that you will practice in this chapter:

- ✓ Writing Polynomials in Standard Form
- ✓ Simplifying Polynomials
- ✓ Adding and Subtracting Polynomials
- ✓ Multiplying Monomials
- ✓ Multiplying and Dividing Monomials
- ✓ Multiplying a Polynomial and a Monomial
- ✓ Multiplying Binomials
- ✓ Factoring Trinomials
- ✓ Operations with Polynomials

Mathematics is the supreme judge; from its decisions there is no appeal. – Tobias Dantzig

Writing Polynomials in Standard Form

✎ **Write each polynomial in standard form.**

1) $9x - 7x =$

2) $-6 + 15x - 15x =$

3) $3x^2 - 11x^3 =$

4) $18 + 19x^3 - 14 =$

5) $3x^2 + 9x - 4x^5 =$

6) $-7x^3 + 12x^7 =$

7) $9x + 6x^2 - 2x^6 =$

8) $-5x^3 + x - 9x^4 =$

9) $8x^2 + 34 - 21x =$

10) $8 - 7x + 11x^4 =$

11) $25x^3 + 45x - 13x^4 =$

12) $17 + 9x^2 - 2x^3 =$

13) $18x^2 - 8x + 8x^3 =$

14) $9x^4 - 4x^2 - 10x^5 =$

15) $-41 + 7x^2 - 8x^4 =$

16) $8x^2 - 7x^5 + 3x^3 - 12 =$

17) $4x^2 - 9x^5 + 12 - 8x^4 =$

18) $-2x^5 + 6x - 9x^2 - 7x =$

19) $14x^5 + 7x^4 - 8x^5 - 8x^2 =$

20) $2x^3 - 15x^4 + 9x^3 + 3x^8 =$

21) $7x^4 - 16x^5 - 9x^2 + 10x^4 =$

22) $5x^2 + 6x^5 + 37x^3 - 9x^5 =$

23) $3x(2x + 5 - 6x^2) =$

24) $12x(x^6 + 2x^3) =$

25) $6x(x^2 + 8x + 4) =$

26) $8x(3 - 2x + 4x^3) =$

27) $7x(2x^3 - 2x^2 + 2) =$

28) $5x(5x^5 + 4x^4 - 1) =$

29) $x(4x^3 + 52x^4 + 2x) =$

30) $6x(3x - 4x^4 + 7x^2) =$

Simplifying Polynomials

✎ **Simplify each expression.**

1) $3(x - 12) =$

2) $5x(2x - 4) =$

3) $7x(5x - 1) =$

4) $6x(3x + 2) =$

5) $5x(2x - 7) =$

6) $9x(x + 8) =$

7) $(3x - 8)(x - 3) =$

8) $(x - 9)(3x + 4) =$

9) $(x - 8)(x - 5) =$

10) $(3x + 4)(3x - 4) =$

11) $(5x - 8)(5x - 2) =$

12) $7x^2 + 7x^2 - 6x^4 =$

13) $5x - 2x^2 + 7x^3 + 10 =$

14) $8x + 2x^2 - 5x^3 =$

15) $15x + 4x^5 - 8x^2 =$

16) $-4x^2 + 7x^5 + 11x^4 =$

17) $-14x^2 + 8x^3 - 2x^4 + 5x =$

18) $14 - 5x^2 + 6x^2 - 10x^3 + 17 =$

19) $x^2 - 9x + 2x^3 + 15x - 10x =$

20) $14 - 8x^2 + 4x^2 - 9x^3 + 1 =$

21) $-4x^5 + 2x^4 - 18x^2 + 2x^5 =$

22) $(3x^3 - 5) + (3x^3 - 2x^3) =$

23) $4(3x^5 - 3x^3 - 6x^5) =$

24) $-4(x^5 + 8) - 4(12 - x^5) =$

25) $7x^2 - 9x^3 - 2x + 14 - 5x^2 =$

26) $10 - 5x^2 + 3x^2 - 4x^3 + 4 =$

27) $(8x^2 - 2x) - (5x - 5 - 4x^2) =$

28) $4x^4 - 8x^3 - x(3x^2 + 5x) =$

29) $4x + 8x^2 - 10 - 2(x^2 - 1) =$

30) $5 - 3x^2 + (6x^4 - 2x^2 + 8x^4) =$

31) $-(x^5 + 8) - 7(4 + x^5) =$

32) $(4x^3 - x) - (x - 6x^3) =$

Adding and Subtracting Polynomials

👌 **Add or subtract expressions.**

1) $(-x^3 - 3) + (4x^3 + 2) =$

2) $(3x^2 + 4) - (6 - x^2) =$

3) $(x^3 + 4x^2) - (5x^3 + 15) =$

4) $(3x^3 - 2x^2) + (2x^2 - x) =$

5) $(10x^3 + 14x) - (14x^3 + 7) =$

6) $(5x^2 - 7) + (3x^2 + 7) =$

7) $(9x^3 + 4) - (10 - 5x^3) =$

8) $(x^2 + 2x^3) - (2x^3 + 5) =$

9) $(8x^2 - x) + (5x - 4x^2) =$

10) $(17x + 10) - (2x + 10) =$

11) $(12x^4 - 4x) - (x - 3x^4) =$

12) $(3x - x^4) - (7x^4 + 8x) =$

13) $(7x^3 - 6x^5) - (4x^5 - 2x) =$

14) $(x^3 - 7) + (4x^3 + 8x^5) =$

15) $(6x^2 + 5x^4) - (x^4 - 9x^2) =$

16) $(-4x^2 - 4x) + (7x - 8x^2) =$

17) $(x - 6x^4) - (15x^4 + 2x) =$

18) $(4x - 3x^4) - (2x^4 - 3x^3) =$

19) $(7x^3 - 7) + (6x^3 - 6x^2) =$

20) $(9x^5 + 7x^4) - (x^4 - 5x^5) =$

21) $(-4x^2 + 11x^4 + 2x^3) + (20x^3 + 4x^4 + 12x^2) =$

22) $(5x^2 - 5x^4 - 5x) - (-4x^2 - 5x^4 + 5x) =$

23) $(12x + 36x^3 - 10x^4) + (20x^3 + 10x^4 - 7x) =$

24) $(2x^5 - 4x^3 - 5x) - (2x^2 + 7x^3 - 2x) =$

25) $(14x^3 - 4x^5 - x) - (-4x^3 - 12x^5 + 9x) =$

26) $(-5x^2 + 12x^4 + x^3) + (10x^3 + 17x^4 + 7x^2) =$

Multiplying Monomials

✏️ **Simplify each expression.**

1) $7u^5 \times (-u^2) =$

2) $(-9p^8) \times (-4p^2) =$

3) $5xy^3z^3 \times 4z^2 =$

4) $8u^6t \times 2ut^2 =$

5) $(-2a^2) \times (-5a^3b^3) =$

6) $-4a^2b^2 \times 5a^4b =$

7) $10xy^4 \times 2x^2y^2 =$

8) $4p^2q^4 \times (-2pq^2) =$

9) $8s^5t^4 \times 3st^4 =$

10) $(-7x^5y^3) \times 7x^4y =$

11) $xy^7z \times 15z^3 =$

12) $15xy \times 2x^3y =$

13) $14pq^4 \times (-3p^3q) =$

14) $25s^4t^2 \times st^6 =$

15) $12p^5 \times (-2p^3) =$

16) $(-12p^2q^4r) \times 3pq^5r^3 =$

17) $(-7a^4) \times (-4a^5b) =$

18) $4u^7v^2 \times (-9u^4v^6) =$

19) $9u^5 \times (-3u) =$

20) $-3xy^9 \times 8x^5y =$

21) $12y^5z^3 \times (-2y^2z) =$

22) $9a^3bc^5 \times 4abc^3 =$

23) $(-9p^5q^2) \times (-3p^2q^4) =$

24) $4u^8v^3 \times (-4u^8v^5) =$

25) $15y^3z^4 \times (-y^5z) =$

26) $(-12pq^4r^3) \times 5p^4q^2r =$

27) $3ab^5c^2 \times 3a^2bc^4 =$

28) $7x^5yz^3 \times 9x^5y^7z^4 =$

Multiplying and Dividing Monomials

✏️ **Simplify each expression.**

1) $(7x^2)(x^3) =$

2) $(4x^3)(5x^2) =$

3) $(3x^4)(2x^2) =$

4) $(5x^8)(8x^3) =$

5) $(12x^6)(2x^3) =$

6) $(2yx^5)(16x^2) =$

7) $(9x^5y)(2x^2y^3) =$

8) $(-2x^2y^5)(5x^3y^2) =$

9) $(-4x^2y^2)(-8x^4y^3) =$

10) $(2x^4y)(-5x^5y^3) =$

11) $(9x^4y^4)(2x^3y^3) =$

12) $(2x^4y^6)(3x^4y^3) =$

13) $(8x^3y^8)(7x^5y^{10}) =$

14) $(14x^6y^5)(3x^5y^5) =$

15) $(8x^2y^8)(5x^{10}y^{10}) =$

16) $(-3x^2y^5)(4x^6y^3) =$

17) $\dfrac{9x^4y^5}{xy^3} =$

18) $\dfrac{18x^8y^3}{18x^7y} =$

19) $\dfrac{54x^4y^4}{6xy} =$

20) $\dfrac{63x^4y^5}{7x^3y^4} =$

21) $\dfrac{32x^7y^6}{8x^2y^3} =$

22) $\dfrac{63x^9y^4}{3x^4y^3} =$

23) $\dfrac{96x^{16}y^{12}}{12x^7y^9} =$

24) $\dfrac{60x^{10}y^6}{12x^{11}y^3} =$

25) $\dfrac{90x^8y^{12}}{18x^7y^{12}} =$

26) $\dfrac{45x^{23}y^{10}}{9x^9y^6} =$

27) $\dfrac{-96x^8y^8}{24x^6y^8} =$

Multiplying a Polynomial and a Monomial

🖎 **Find each product.**

1) $2x(x + 4) =$

2) $3(8 - x) =$

3) $5x(3x + 4) =$

4) $x(-2x + 5) =$

5) $7x(3x - 3) =$

6) $3(2x - 5y) =$

7) $6x(7x - 3) =$

8) $x(12x + 5y) =$

9) $5x(x + 6y) =$

10) $11x(4x + 5y) =$

11) $8x(4x + 2) =$

12) $12x(x - 15y) =$

13) $9x(5x - 3y) =$

14) $8x(5x - 2y + 5) =$

15) $9x(2x^2 + 7y^2) =$

16) $8x(9x + 6y) =$

17) $2(3x^5 - 2y^5) =$

18) $4x(-x^2y + 2y) =$

19) $-3(2x^3 - 3xy + 9) =$

20) $2(x^2 - 2xy - 4) =$

21) $7x(4x^3 - xy + 2x) =$

22) $-9x(-2x^3 - 2x + 7xy) =$

23) $6(x^2 + 3xy - 8y^2) =$

24) $5x(7x^3 - x + 8) =$

25) $7(x^{24} - 4x - 6) =$

26) $x^2(-3x^3 + 4x + 7) =$

27) $x^2(2x^3 + 10 - 5x) =$

28) $4x^4(3x^3 - 2x + 8) =$

29) $5x^2(x^4 - 5xy + 2y^3) =$

30) $4x^2(7x^4 - 2x + 11) =$

31) $7x^3(3x^3 + 5x - 7) =$

32) $4x(x^2 - 8xy + 7y^3) =$

Pre-Algebra Workbook

Multiplying Binomials

✎ **Find each product.**

1) $(x+5)(x+1) =$

2) $(x-3)(x+7) =$

3) $(x-1)(x-9) =$

4) $(x+3)(x+8) =$

5) $(x-4)(x-11) =$

6) $(x+5)(x+6) =$

7) $(x-8)(x+7) =$

8) $(x-3)(x-2) =$

9) $(x+8)(x+11) =$

10) $(x-3)(x+5) =$

11) $(x+8)(x+8) =$

12) $(x+2)(x+7) =$

13) $(x-9)(x+4) =$

14) $(x-10)(x+10) =$

15) $(x+24)(x+2) =$

16) $(x+9)(x+13) =$

17) $(x-7)(x+7) =$

18) $(x-5)(x+2) =$

19) $(3x+4)(x+5) =$

20) $(x-8)(5x+2) =$

21) $(x-9)(4x+9) =$

22) $(2x-7)(3x-2) =$

23) $(x-4)(x+11) =$

24) $(5x-6)(2x+4) =$

25) $(4x-9)(x+7) =$

26) $(8x-5)(2x+2) =$

27) $(3x+9)(7x+4) =$

28) $(6x-8)(4x+4) =$

29) $(4x+5)(5x-8) =$

30) $(8x-1)(8x+4) =$

31) $(9x+4)(3x-6) =$

32) $(4x^2+12)(4x^2-12) =$

Factoring Trinomials

✎ **Factor each trinomial.**

1) $x^2 + 12x + 35 =$

2) $x^2 - 8x + 12 =$

3) $x^2 + 11x + 10 =$

4) $x^2 - 12x + 27 =$

5) $x^2 - 16x + 15 =$

6) $x^2 - 13x + 40 =$

7) $x^2 + 15x + 44 =$

8) $x^2 + x - 72 =$

9) $x^2 - 81 =$

10) $x^2 - 17x + 70 =$

11) $x^2 + 8x - 48 =$

12) $x^2 + 5x - 104 =$

13) $x^2 - 7x - 18 =$

14) $x^2 + 22x + 121 =$

15) $3x^2 - 3x - 36 =$

16) $2x^2 - 35x + 75 =$

17) $14x^2 + 11x - 15 =$

18) $8x^2 - 12x - 20 =$

19) $15x^2 + 16x + 4 =$

20) $24x^2 + 2x - 1 =$

✎ **Calculate each problem.**

21) The area of a rectangle is $x^2 - 3x - 40$. If the width of rectangle is $x - 8$, what is its length? _____

22) The area of a parallelogram is $12x^2 + 7x - 10$ and its height is $4x + 5$. What is the base of the parallelogram? _____

23) The area of a rectangle is $10x^2 - 43x + 28$. If the width of the rectangle is $5x - 4$, what is its length? _____

Operations with Polynomials

🖎 **Find each product.**

1) $2(4x + 1) =$ _____

2) $5(2x + 7) =$ _____

3) $4(6x - 5) =$ _____

4) $-4(7x - 8) =$ _____

5) $3x^2(8x + 4) =$ _____

6) $6x^2(2x - 9) =$ _____

7) $5x^3(-x + 4) =$ _____

8) $-5x^4(4x - 9) =$ _____

9) $6(x^2 + 7x - 3) =$ _____

10) $4(3x^2 - 2x + 6) =$ _____

11) $9(3x^2 + 8x + 2) =$ _____

12) $7x(x^2 + 5x + 3) =$ _____

13) $(7x + 2)(2x - 5) =$ _____

14) $(8x + 5)(3x - 8) =$ _____

15) $(4x + 2)(6x - 1) =$ _____

16) $(5x - 4)(5x + 9) =$ _____

🖎 **Calculate each problem.**

17) The measures of two sides of a triangle are $(2x + 8y)$ and $(5x - 3y)$. If the perimeter of the triangle is $(11x + 6y)$, what is the measure of the third side? _____

18) The height of a triangle is $(8x + 2)$ and its base is $(2x - 6)$. What is the area of the triangle? _____

19) One side of a square is $(4x + 3)$. What is the area of the square? _____

20) The length of a rectangle is $(7x - 9y)$ and its width is $(13x + 9y)$. What is the perimeter of the rectangle? _____

21) The side of a cube measures $(x + 2)$. What is the volume of the cube? _____

22) If the perimeter of a rectangle is $(24x + 10y)$ and its width is $(4x + 3y)$, what is the length of the rectangle? _____

Answers of Worksheets – Chapter 7

Writing Polynomials in Standard Form

1) $2x$
2) -6
3) $-11x^3 + 3x^2$
4) $19x^4 + 4$
5) $-4x^5 + 3x^2 + 9x$
6) $12x^7 - 7x^3$
7) $-2x^6 + 6x^2 + 9x$
8) $-9x^4 - 5x^3 + x$
9) $8x^2 - 21x + 34$
10) $11x^4 - 7x + 8$
11) $-13x^4 + 25x^3 + 45x$
12) $-2x^3 + 9x^2 + 17$
13) $8x^3 + 18x^2 - 8x$
14) $-10x^5 - 9x^4 - 4x^2$
15) $-8x^4 + 7x^2 - 41$
16) $-7x^5 + 3x^3 + 8x^2 - 12$
17) $-9x^5 - 8x^4 + 4x^2 + 12$
18) $-2x^5 - 9x^2 - x$
19) $6x^5 + 7x^4 - 8x^2$
20) $3x^8 - 15x^4 + 11x^2$
21) $-16x^5 + 17x^4 - 9x^2$
22) $-3x^5 + 37x^3 + 5x^2$
23) $-18x^3 + 6x^2 + 15x$
24) $12x^7 + 24x^4$
25) $6x^3 + 48x^2 + 24x$
26) $32x^4 - 16x^2 + 24x$
27) $14x^4 - 14x^3 + 14x$
28) $25x^6 + 20x^5 - 5x$
29) $52x^5 + 4x^4 + 2x^2$
30) $-24x^5 + 42x^3 + 18x^2$

Simplifying Polynomials

1) $3x - 36$
2) $10x^2 - 20x$
3) $35x^2 - 7x$
4) $18x^2 + 12x$
5) $10x^2 - 35x$
6) $9x^2 + 72x$
7) $3x^2 - 17x + 24$
8) $3x^2 - 23x - 36$
9) $x^2 - 13x + 40$
10) $9x^2 - 16$
11) $25x^2 - 50x + 16$
12) $-6x^4 + 14x^2$
13) $7x^3 - 2x^2 + 5x + 10$
14) $-5x^3 + 2x^2 + 8x$
15) $4x^5 - 8x^2 + 15x$
16) $7x^5 + 11x^4 - 4x^2$
17) $-2x^4 + 8x^3 - 14x^2 + 5x$
18) $-10x^3 + x^2 + 31$
19) $2x^3 + x^2 - 4x$
20) $-9x^3 - 4x^2 + 15$
21) $-2x^5 + 2x^4 - 18x^2$
22) $4x^3 - 5$
23) $-12x^5 - 12x^3$
24) -80

25) $-9x^3 + 2x^2 - 2x + 14$

26) $-4x^3 - 2x^2 + 14$

27) $12x^2 - 7x + 5$

28) $4x^4 - 11x^3 - 5x^2$

29) $6x^2 + 4x - 8$

30) $14x^4 - 5x^2 + 5$

31) $-8x^5 - 36$

32) $10x^3 - 2x$

Adding and Subtracting Polynomials

1) $3x^2 - 1$

2) $4x^2 - 2$

3) $-4x^3 + 4x^2 - 15$

4) $3x^3 - x$

5) $-4x^3 + 14x - 7$

6) $8x^2$

7) $14x^3 - 6$

8) $x^2 - 5$

9) $4x^2 + 4x$

10) $15x$

11) $15x^4 - 5x$

12) $-8x^4 - 5x$

13) $-10x^5 + 7x^3 + 2x$

14) $5x^5 + 5x^3 - 7$

15) $4x^4 + 15x^2$

16) $-12x^2 + 3x$

17) $-21x^4 - x$

18) $-5x^4 + 3x^3 + 4x$

19) $13x^3 - 6x^2 - 7$

20) $14x^5 + 6x^4$

21) $15x^4 + 22x^3 + 8x^2$

22) $9x^2 - 10x$

23) $56x^3 + 5x$

24) $2x^5 - 11x^3 - 2x^2 - 3x$

25) $8x^5 + 18x^3 - 10x$

26) $29x^4 + 11x^3 + 2x^2$

Multiplying Monomials

1) $-7u^7$

2) $36p^{10}$

3) $20xy^3z^5$

4) $16u^7t^3$

5) $10a^5b^3$

6) $-20a^6b^3$

7) $20x^3y^6$

8) $-8p^3q^6$

9) $24s^6t^8$

10) $-49x^9y^4$

11) $15xy^7z^4$

12) $30x^4y^2$

13) $-42p^4q^5$

14) $25s^5t^8$

15) $-24p^8$

16) $-36p^3q^9r^4$

17) $28a^9b$

18) $-36u^{11}v^8$

19) $-27u^6$

20) $-24x^6y^{10}$

21) $-24y^7z^4$

22) $36a^4b^2c^8$

23) $27p^7q^6$

24) $-16u^{16}v^8$

25) $-15y^8z^5$

26) $-60p^5q^6r^4$

27) $9a^3b^6c^6$

28) $63x^{10}y^8z^7$

Multiplying and Dividing Monomials

1) $7x^5$

2) $20x^5$

3) $6x^6$

4) $40x^{11}$

5) $24x^9$

6) $32x^7y$

7) $18x^7y^4$

8) $-10x^5y^7$

9) $32x^6y^5$

10) $-10x^9y^4$

11) $18x^7y^7$

12) $6x^8y^9$

13) $56x^8y^{18}$
14) $42x^{11}y^{10}$
15) $40x^{12}y^{18}$
16) $-12x^8y^8$
17) $9x^3y^2$

18) xy^2
19) $9x^3y^3$
20) $9xy$
21) $4x^5y^3$
22) $21x^5y$

23) $8x^9y^3$
24) $5x^{-1}y^3$
25) $5x$
26) $5x^{14}y^4$
27) $-4x^2$

Multiplying a Polynomial and a Monomial

1) $2x^2 + 8x$
2) $-3x + 24$
3) $15x^2 + 20x$
4) $-2x^2 + 5x$
5) $21x^2 - 21x$
6) $6x - 15y$
7) $42x^2 - 18x$
8) $12x^2 + 5xy$
9) $5x^2 + 30xy$
10) $44x^2 + 55xy$
11) $32x^2 + 16x$
12) $12 - 180xy$
13) $45x^2 - 27xy$
14) $40x^2 - 16xy + 40x$
15) $18x^3 + 63xy^2$
16) $72x^2 + 48xy$

17) $6x^5 - 2y^5$
18) $-4x^3y + 8xy$
19) $-6x^3 + 9xy - 27$
20) $2x^2 - 4xy - 8$
21) $28x^4 - 7x^2y + 14x^2$
22) $18x^4 + 18x^2 - 63x^2y$
23) $6x^2 + 18xy - 48y^2$
24) $35x^4 - 5x^2 + 40x$
25) $7x^{24} - 28x - 42$
26) $-3x^5 + 4x^3 + 7x^2$
27) $2x^5 - 5x^3 + 10x^2$
28) $12x^7 - 8x^5 + 32x^4$
29) $5x^6 - 25x^3y + 10x^2y^3$
30) $28x^6 - 8x^3 + 44x^2$
31) $21x^6 + 35x^4 - 49x^3$
32) $4x^3 - 32x^2y + 28xy^3$

Multiplying Binomials

1) $x^2 + 6x + 5$
2) $x^2 + 4x - 21$
3) $x^2 - 10x + 9$
4) $x^2 + 11x + 24$
5) $x^2 - 15x + 44$
6) $x^2 + 11x + 30$
7) $x^2 - x - 56$

8) $x^2 - 5x + 6$
9) $x^2 + 19x + 88$
10) $x^2 + 2x + 15$
11) $x^2 + 16x + 64$
12) $x^2 + 9x + 14$
13) $x^2 - 5x - 36$
14) $x^2 - 100$

15) $x^2 + 26x + 48$
16) $x^2 + 22x + 117$
17) $x^2 - 49$
18) $x^2 - 3x - 10$
19) $3x^2 + 19x + 20$
20) $5x^2 - 38x - 16$
21) $4x^2 - 27x - 81$
22) $6x^2 - 25x + 14$
23) $x^2 + 7x - 44$

24) $10x^2 + 8x - 24$
25) $4x^2 + 19x - 63$
26) $16x^2 + 6x - 10$
27) $21x^2 + 75x + 36$
28) $24x^2 - 8x - 32$
29) $20x^2 - 7x - 40$
30) $64x^2 + 24x - 4$
31) $27x^2 - 42x - 24$
32) $16x^4 - 144$

Factoring Trinomials

1) $(x+5)(x+7)$
2) $(x-2)(x-6)$
3) $(x+1)(x+10)$
4) $(x-9)(x-3)$
5) $(x-1)(x-15)$
6) $(x-5)(x-8)$
7) $(x+4)(x+11)$
8) $(x+9)(x-8)$

9) $(x-9)(x+9)$
10) $(x-7)(x-10)$
11) $(x-4)(x+12)$
12) $(x-8)(x+13)$
13) $(x+2)(x-9)$
14) $(x+11)(x+11)$
15) $(3x+9)(x-4)$
16) $(x-15)(2x-5)$

17) $(7x-5)(2x+3)$
18) $(2x-5)(4x+4)$
19) $(3x+2)(5x+2)$
20) $(6x-1)(4x+1)$
21) $(x+5)$
22) $(3x-2)$
23) $(2x-7)$

Operations with Polynomials

1) $8x + 2$
2) $10x + 35$
3) $24x - 20$
4) $-28x + 32$
5) $24x^3 + 12x^2$
6) $12x^3 - 54x^2$
7) $-5x^4 + 20x^3$
8) $-20x^5 + 45x^4$

9) $6x^2 + 42x - 18$
10) $12x^2 - 8x + 24$
11) $27x^2 + 72x + 18$
12) $7x^3 + 35x^2 + 21x$
13) $14x^2 - 31x - 10$
14) $24x^2 - 49x - 40$
15) $24x^2 + 8x - 2$
16) $25x^2 + 25x - 36$

17) $(4x + y)$
18) $8x^2 - 22x - 6$
19) $16x^2 + 24x + 9$
20) $40x$
21) $x^3 + 6x^2 + 12x + 8$
22) $(8x + 2y)$

Chapter 8:
Geometry and Solid Figures

Topics that you will practice in this chapter:

- ✓ Angles
- ✓ Pythagorean Relationship
- ✓ Triangles
- ✓ Polygons
- ✓ Trapezoids
- ✓ Circles
- ✓ Cubes
- ✓ Rectangular Prism
- ✓ Cylinder
- ✓ Pyramids and Cone

Mathematics is, as it were, a sensuous logic, and relates to philosophy as do the arts, music, and plastic art to poetry. — *K. Shegel*

Angles

✎ **What is the value of *x* in the following figures?**

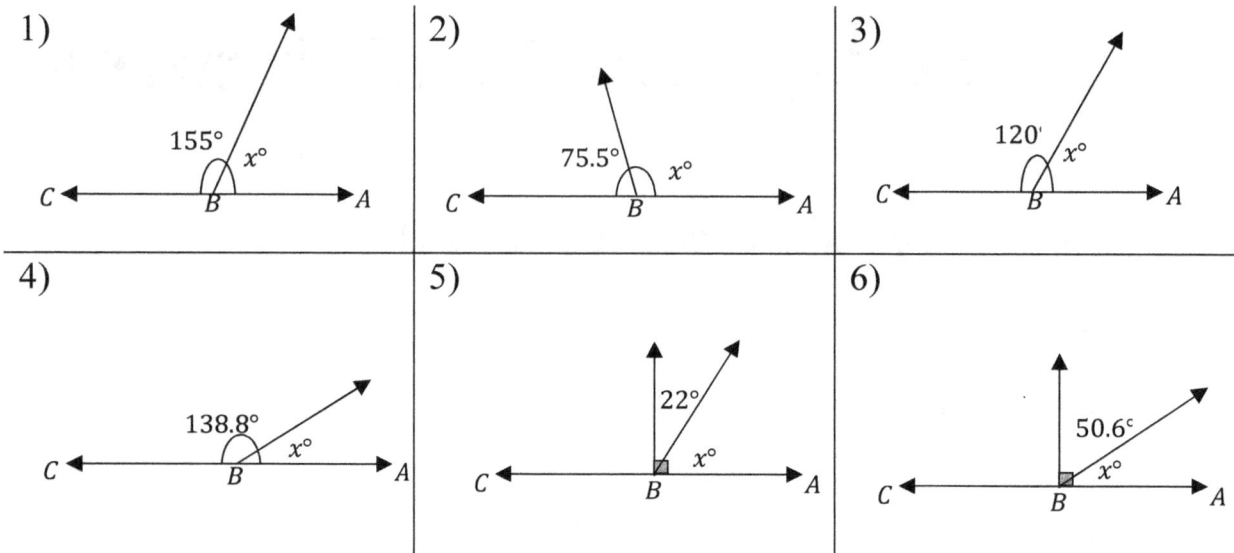

✎ **Calculate.**

7) Two supplementary angles have equal measures. What is the measure of each angle? _____

8) The measure of an angle is nine seventh the measure of its supplement. What is the measure of the angle? _____

9) Two angles are complementary and the measure of one angle is 24 less than the other. What is the measure of the bigger angle? _____

10) Two angles are complementary. The measure of one angle is one fifth the measure of the other. What is the measure of the smaller angle? _____

11) Two supplementary angles are given. The measure of one angle is 80° less than the measure of the other. What does the bigger angle measure? _____

Pythagorean Relationship

✎ Do the following lengths form a right triangle?

1)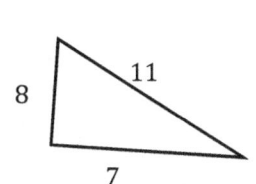

2)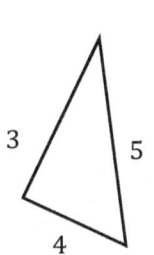

3)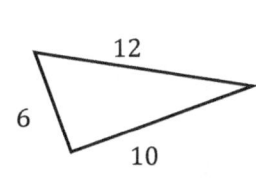

4)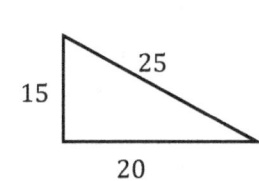

5)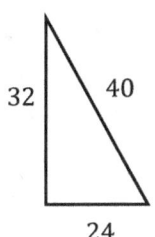

6)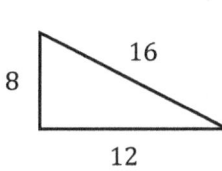

7)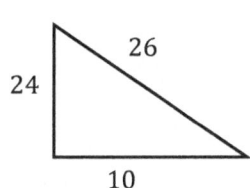

8)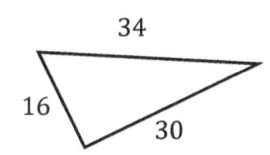

✎ Find the missing side?

9)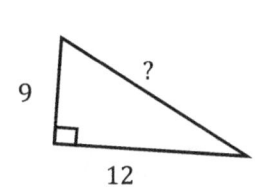

10)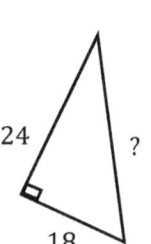

11)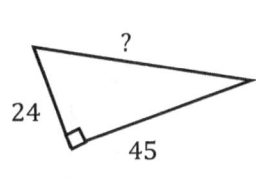

12)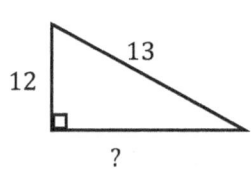

13)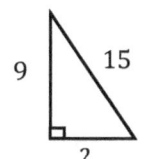

14)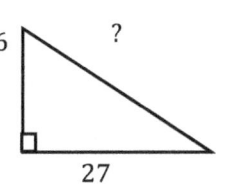

15)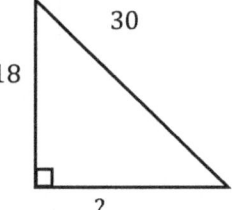

16)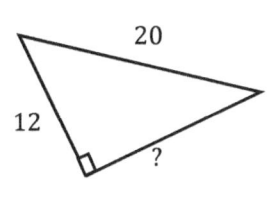

Triangles

✏️ **Find the measure of the unknown angle in each triangle.**

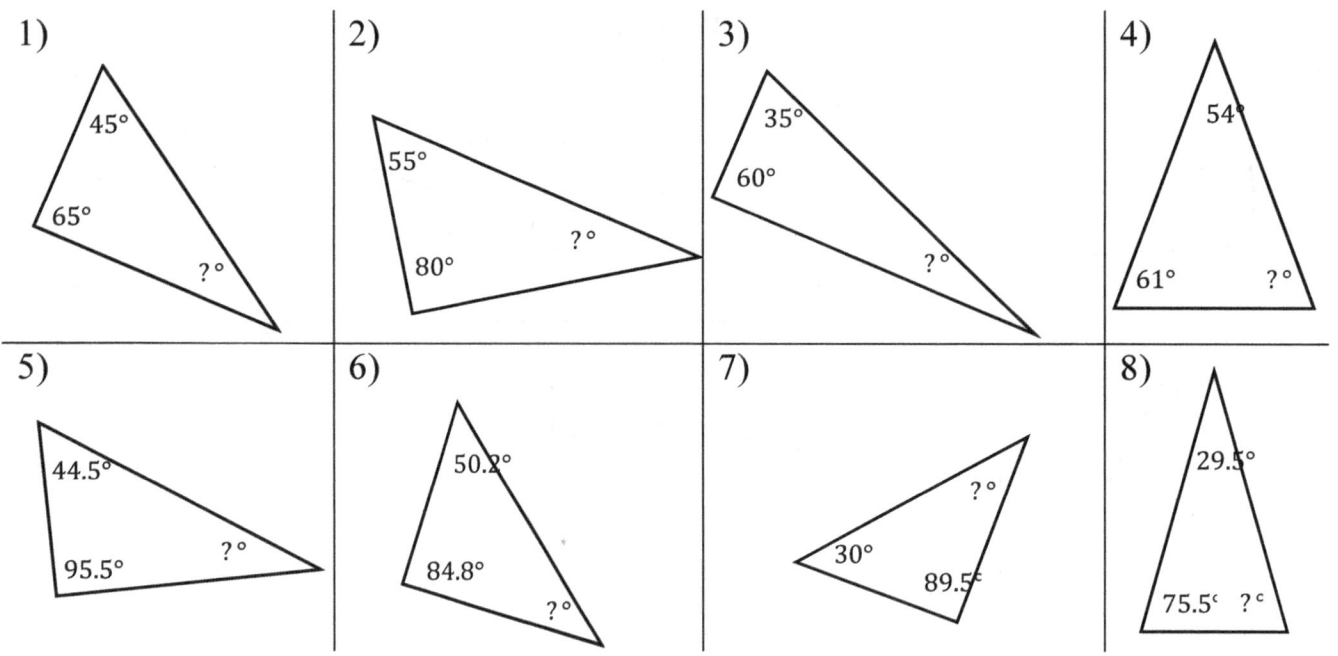

✏️ **Find area of each triangle.**

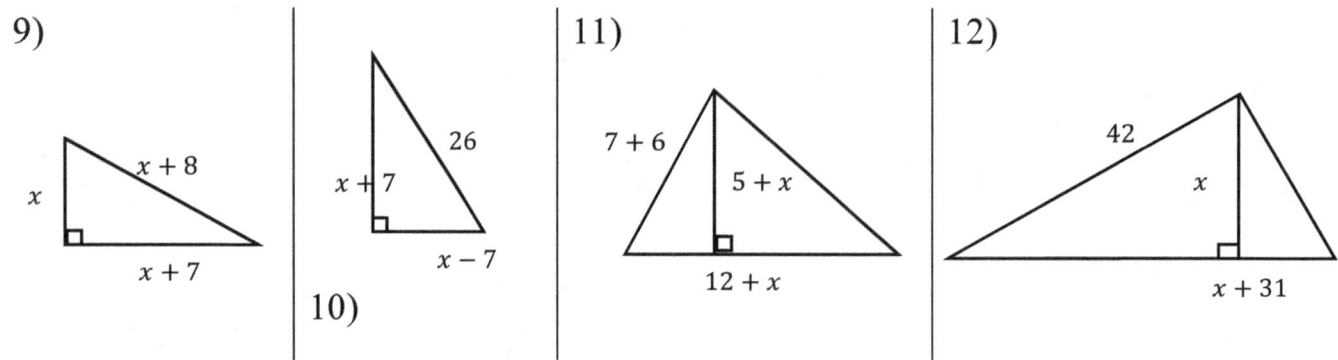

Polygons

✎ **Find the perimeter of each shape.**

1)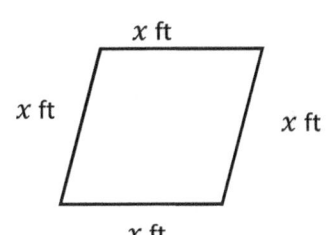

2)
x +2

x in   x in

x+2

3)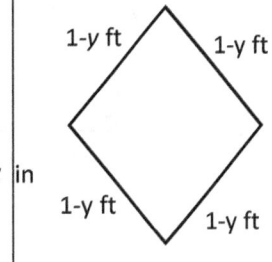

4) Square

(x + 1) cm

5) Regular hexagon

(2x) m

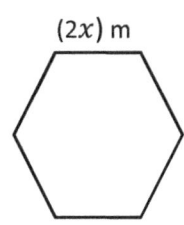

6)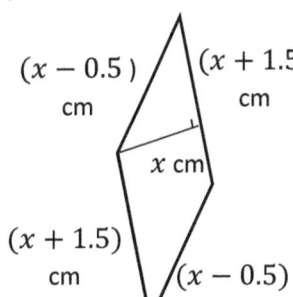

(x − 0.5) cm (x + 1.5) cm

x cm

(x + 1.5) cm (x − 0.5) cm

7) Parallelogram

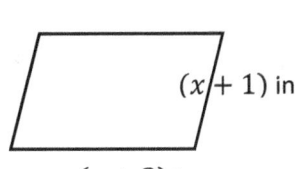

(x + 1) in

(x + 3) in

8) Square

(x + 2) m

✎ **Find the area of each shape.**

9) Parallelogram

2x m

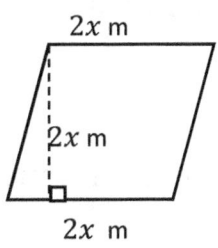

2x m

2x m

10) Rectangle

$10x\ m^2$

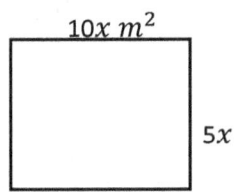

5x

11) Rectangle

(2+x) km

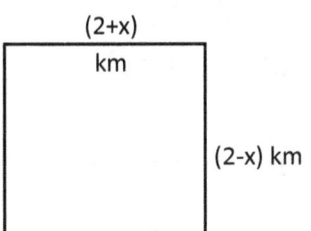

(2-x) km

12) Square

0.6x m

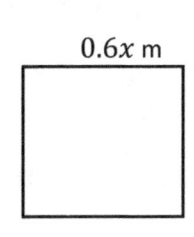

WWW.MathNotion.Com

Trapezoids

🖋 **Find the area of each trapezoid.**

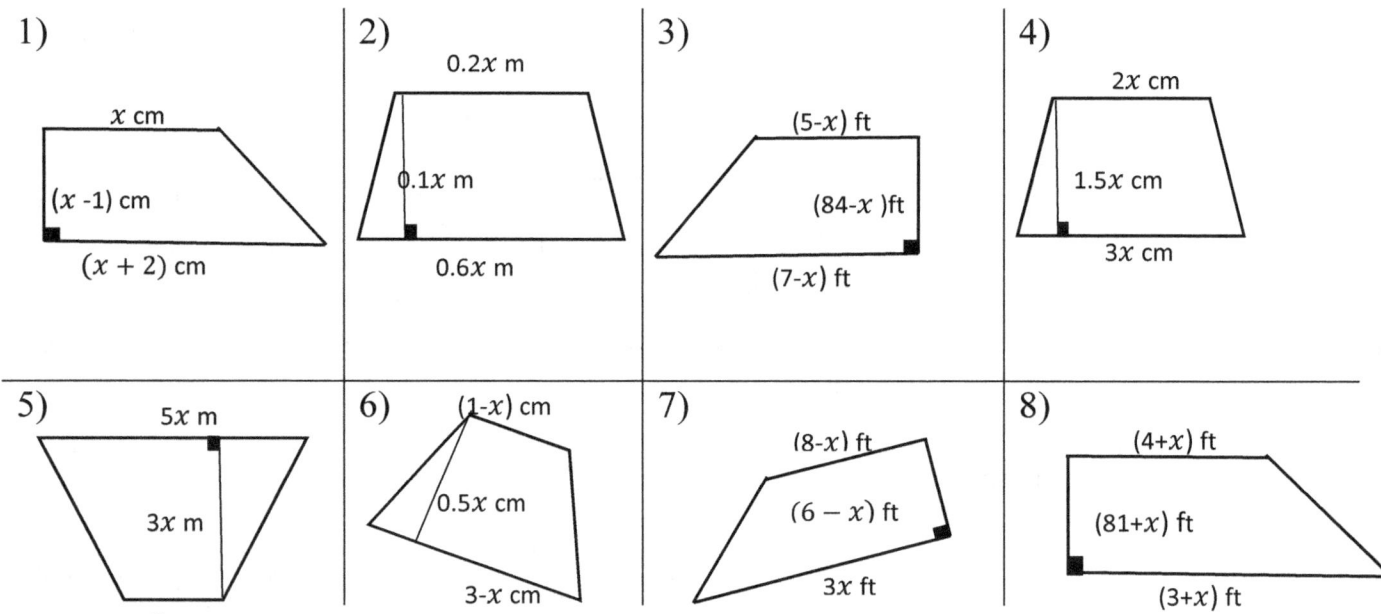

🖋 **Calculate.**

1) A trapezoid has an area of 40 cm² and its height is 8 cm and one base is 6 cm. What is the other base length? _____

2) If a trapezoid has an area of 85 ft² and the lengths of the bases are 9 ft and 8 ft, find the height. _____

3) If a trapezoid has an area of 150 m² and its height is 15 m and one base is 9 m, find the other base length. _____

4) The area of a trapezoid is 196 ft² and its height is 14 ft. If one base of the trapezoid is 12 ft, what is the other base length? _____

Circles

🔖 **Find the area of each circle.** ($\pi = 3.14$)

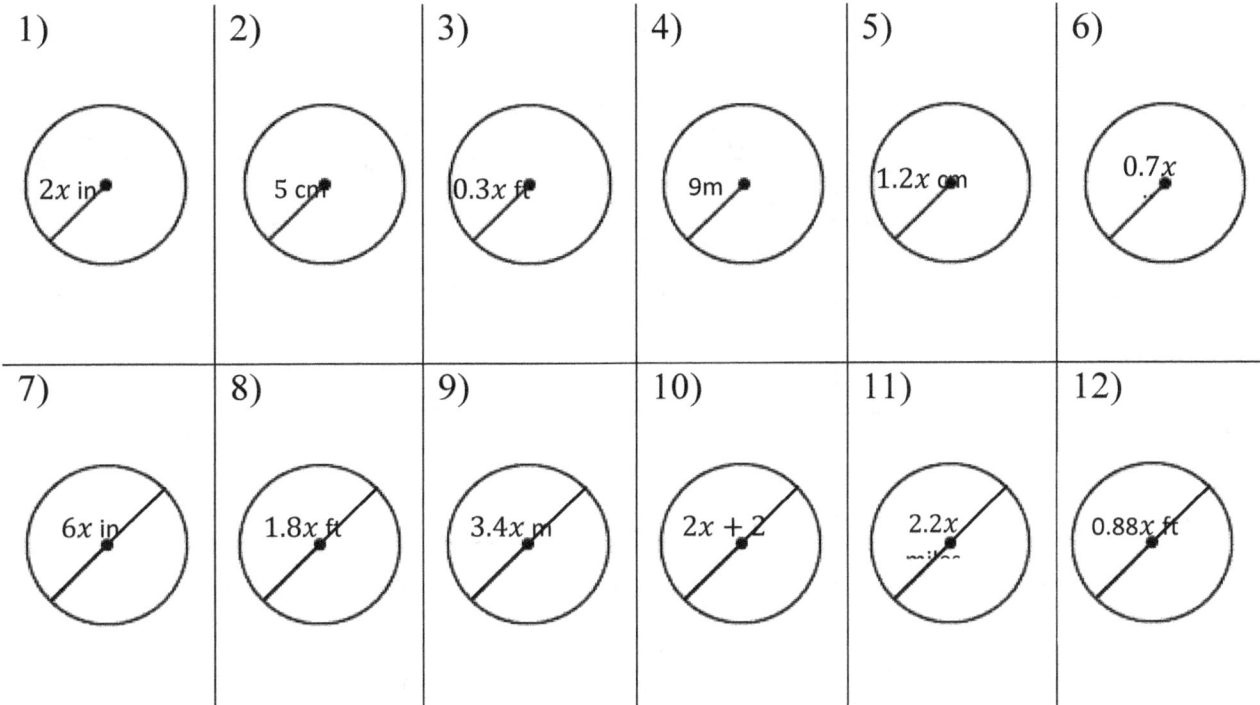

1) 2x in (radius)
2) 5 cm (radius)
3) 0.3x ft (radius)
4) 9m (radius)
5) 1.2x cm (radius)
6) 0.7x (radius)
7) 6x in (diameter)
8) 1.8x ft (diameter)
9) 3.4x m (diameter)
10) 2x + 2 (diameter)
11) 2.2x miles (diameter)
12) 0.88x ft (diameter)

🔖 **Complete the table below.** ($\pi = 3.14$)

Circle No.	Radius	Diameter	Circumference	Area
1	1.4 inches	2.8 inches	8.792 inches	6.154 square inches
2		4.6 meters		
3				$2.01x^2$ square ft
4			36.42 miles	
5		6.2x kilometers		
6	5x centimeters			
7		2x feet		
8				1.54 square meters
9			5.7x inches	
10	(1-x) feet			

Cubes

✎ **Find the volume of each cube.**

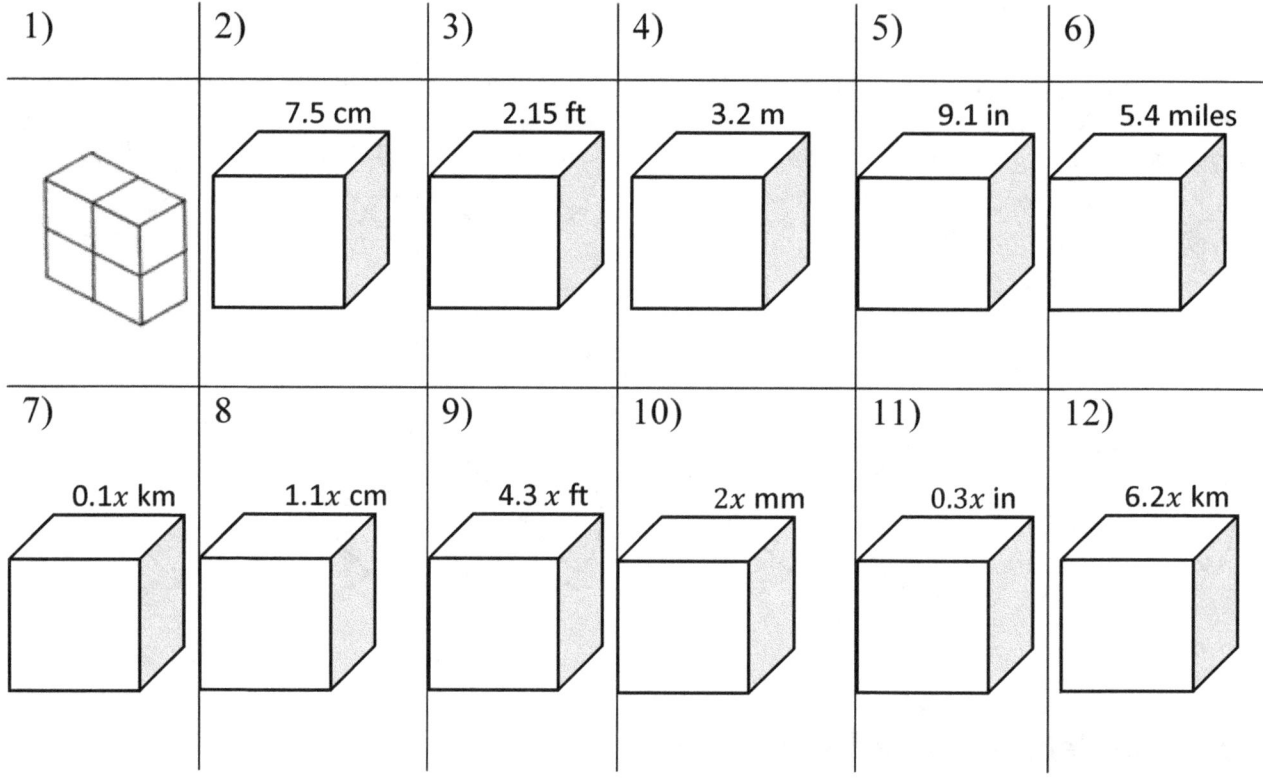

✎ **Find the surface area of each cube.**

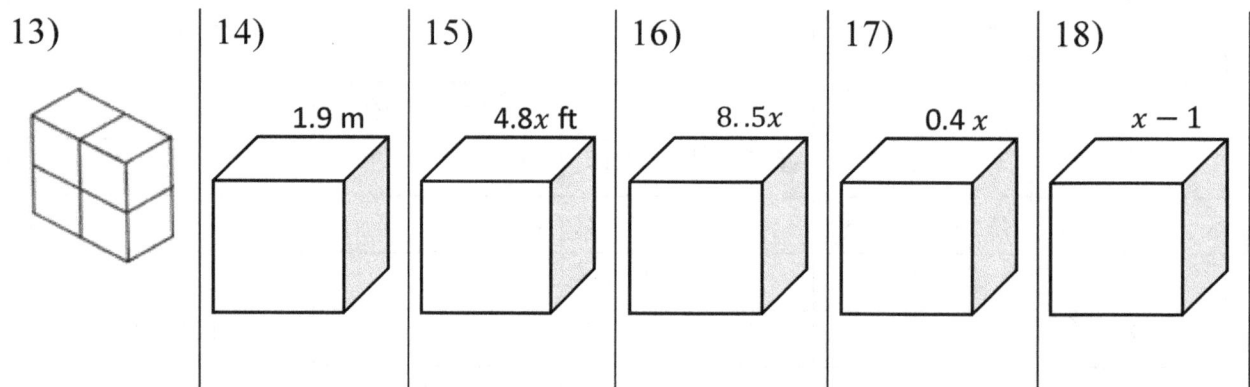

Rectangular Prism

✏️ **Find the volume of each Rectangular Prism.**

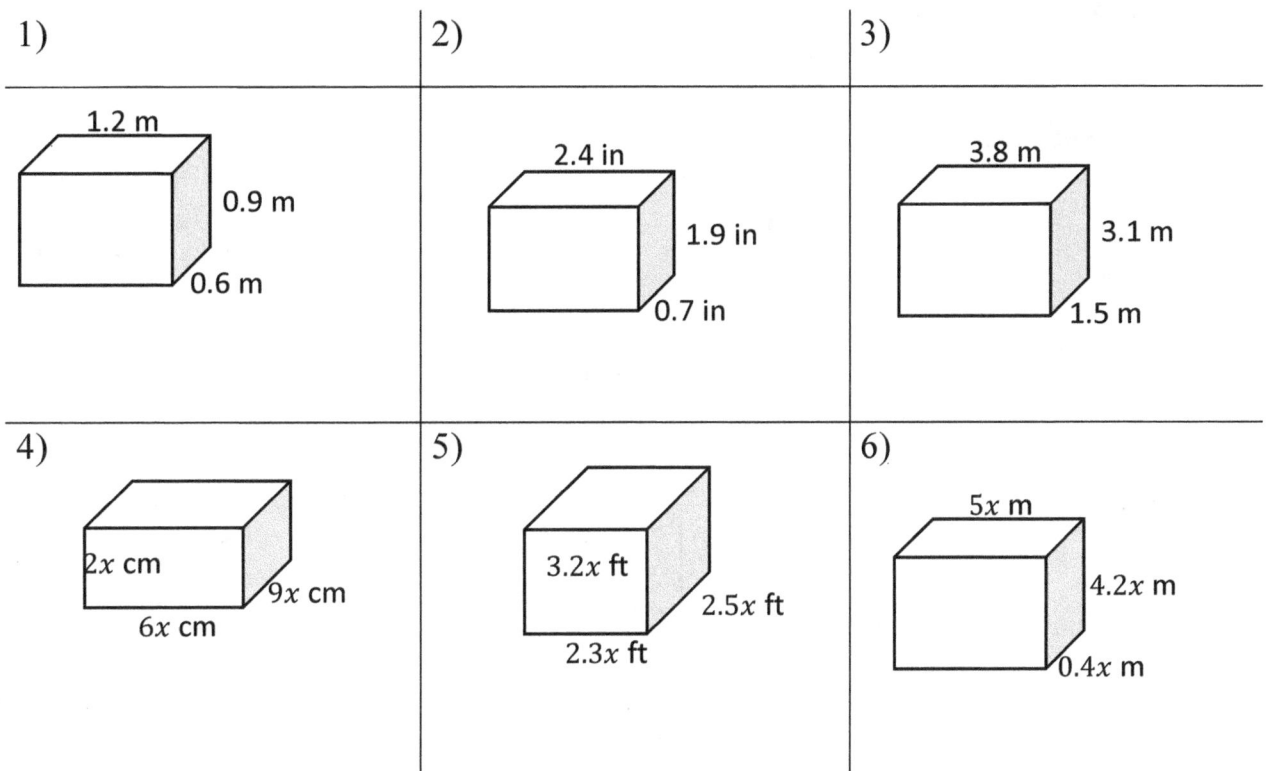

✏️ **Find the surface area of each Rectangular Prism.**

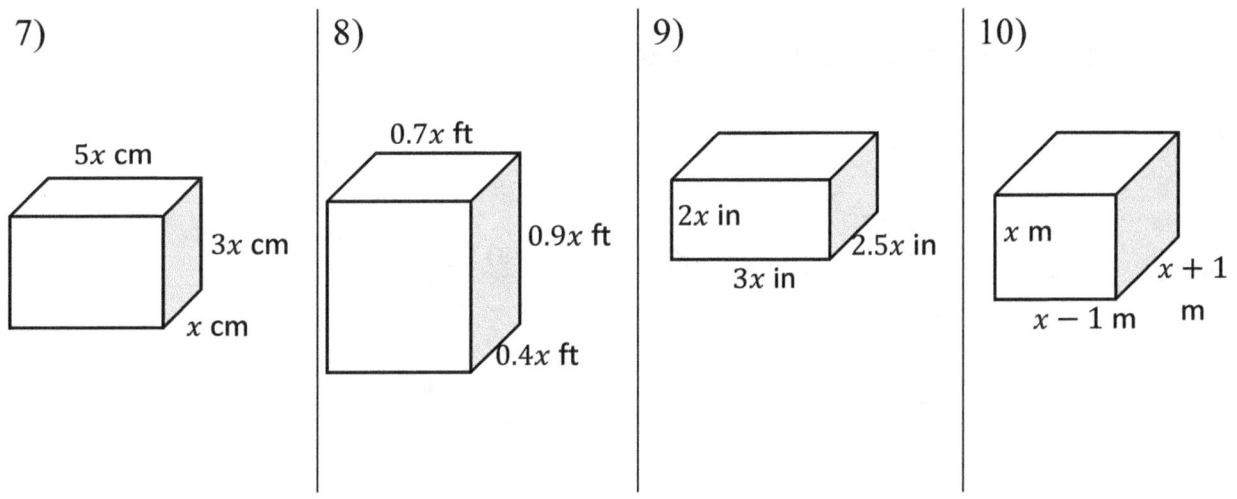

Pre-Algebra Workbook

Cylinder

✎ **Find the volume of each Cylinder. Round your answer to the nearest tenth.** ($\pi = 3.14$)

1)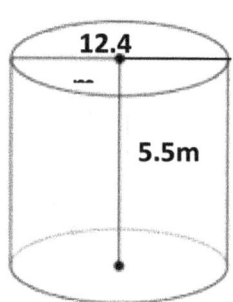

2)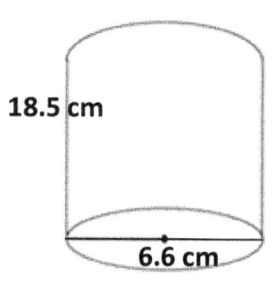

3)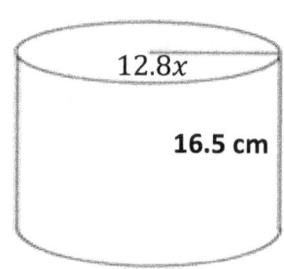

4)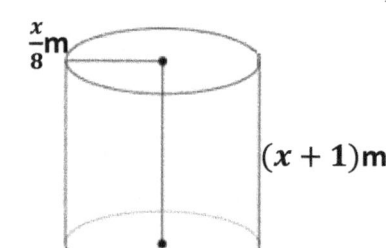

5)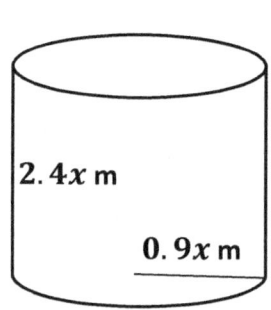

6)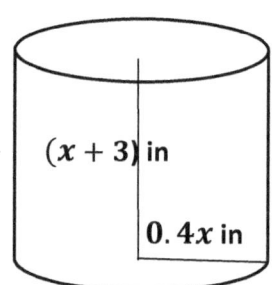

✎ **Find the surface area of each Cylinder.** ($\pi = 3.14$)

7)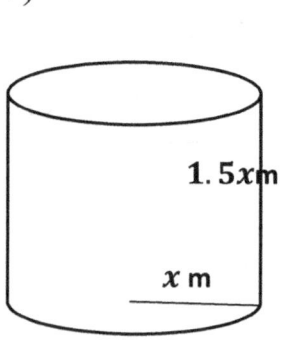

8)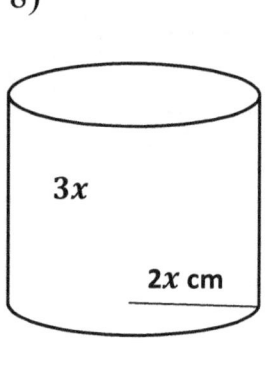

9)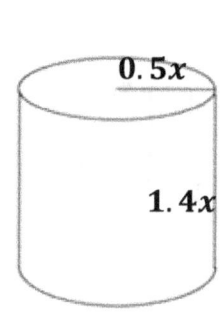

10)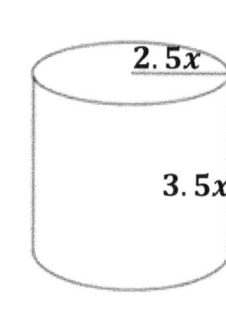

WWW.MathNotion.Com

Pre-Algebra Workbook

Pyramids and Cone

✏️ **Find the volume of each Pyramid and Cone.** ($\pi = 3.14$)

1)
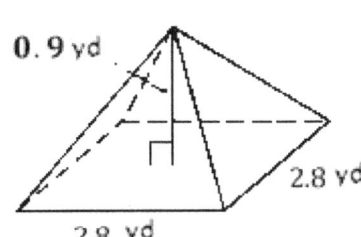

2)
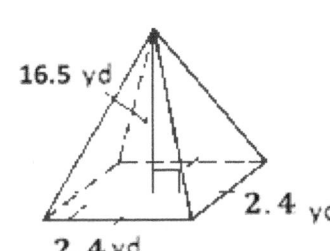

3)
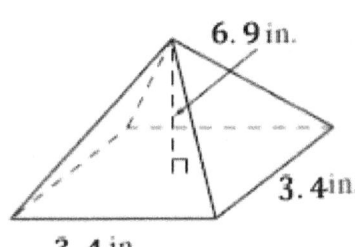

4)
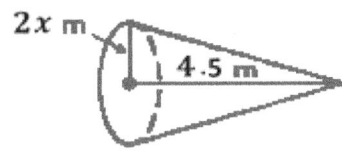

5)
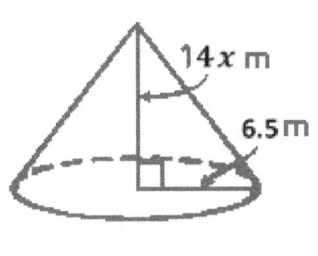

6)
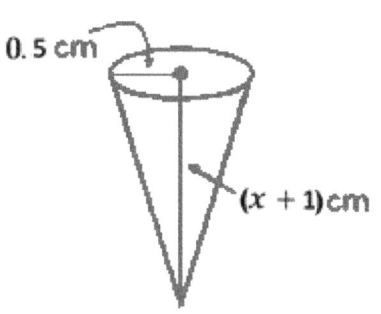

✏️ **Find the surface area of each Pyramid and Cone.** ($\pi = 3.14$)

7)
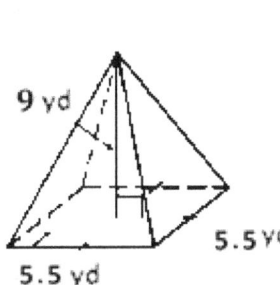

8)
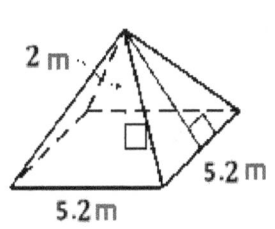

9)
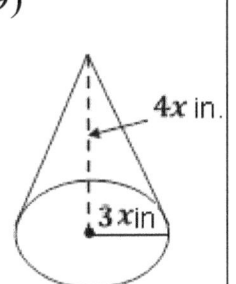

10)
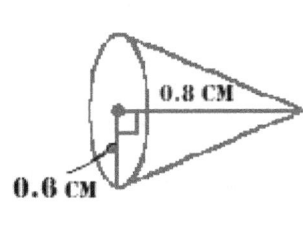

WWW.MathNotion.Com

Answers of Worksheets – Chapter 8

Angles

1) 25°
2) 104.5°
3) 60°
4) 41.2°
5) 68°
6) 39.4°
7) 90°
8) 101.25°
9) 57°
10) 15°
11) 130°

Pythagorean Relationship

1) No
2) Yes
3) No
4) Yes
5) Yes
6) No
7) Yes
8) Yes
9) 15
10) 30
11) 51
12) 5
13) 12
14) 45
15) 24
16) 16

Triangles

1) 70°
2) 45°
3) 85°
4) 65°
5) 40°
6) 45°
7) 60.5°
8) 75°
9) $(\frac{x^2+7x}{2})$ square unites
10) $(\frac{x^2-49}{2})$ square unites
11) $(\frac{x^2+17x+60}{2})$ square unites
12) $(\frac{x^2+31x}{2})$ square unites

Polygons

1) $(4x)\, ft$
2) $(4x+4)\, in$
3) $(4-4y)\, ft$
4) $(4x+4)\, cm$
5) $(12x)\, m$
6) $(4x+2)\, cm$
7) $(4x+8)\, in$
8) $(4x+8)\, m$
9) $(4x^2)m^2$
10) $(50x^2)m^2$
11) $(4-x^2)\, km^2$
12) $(0.36x^2)\, m^2$

Trapezoids

1) $(x^2-1)\, cm^2$
2) $(0.04x^2)\, m^2$
3) $(x^2-10x+24)\, ft^2$
4) $(3.75x^2)\, cm^2$
5) $(10.5x^2)m^2$
6) $(x-0.5x^2)cm^2$
7) $(-x^2+2x+24)\, ft^2$
8) $(\frac{2x^2+9x+7}{2})ft^2$

Calculate

1) 4 cm
2) 10 ft
3) 11 m
4) 16 ft

Circles

1) $(12.56x^2)\, in^2$
2) $78.5\, cm^2$
3) $(0.283x^2)\, ft^2$
4) $254.34 m^2$
5) $(4.522x^2)cm^2$
6) $(1.54x^2)\, miles^2$
7) $(28.56x^2)\, in^2$
8) $(2.543x^2)ft^2$
9) $(9.075x^2)\, m^2$

Pre-Algebra Workbook

10) $(3.14x^2 + 6.28x + 3.14)\ cm^2$ 11) $(3.8x^2)\ miles^2$ 12) $(0.608x^2)\ ft^2$

Circle No.	Radius	Diameter	Circumference	Area
1	1.4 inches	2.8 inches	8.792 inches	6.154 square inches
2	2.3 meters	4.6 meters	14.44 meters	16.61 meters
3	$0.8x$ square ft	$1.6x$ square ft	$5.024x$ square ft	$2.01x^2$ square ft
4	5.8 miles	11.6 miles	36.42 miles	105.63 miles
5	$3.1x$ kilometers	$6.2x$ kilometers	$19.47x$ kilometers	$30.175x^2$ kilometers
6	$5x$ centimeters	$10x$ centimeters	$31.4x$ centimeters	$78.5x^2$ centimeters
7	x feet	$2x$ feet	$6.28x$ feet	$3.14x^2$ feet
8	0.7 square meters	1.4 square meters	4.396 square meters	1.54 square meters
9	$2.5x$ inches	$5x$ inches	$15.7x$ inches	$19.625x^2$ inches
10	$(1-x)$ feet	$2 - 2x$ feet	$6.28 - 6.28x)$ feet	$3.14x^2 - 6.28x + 3.14)$ feet

Cubes
1) 4
2) $421.88\ cm^3$
3) $9.94\ ft^3$
4) $32.77\ m^3$
5) $753.57\ in^3$
6) $157.46\ miles^3$
7) $(0.001x^3)\ km^3$
8) $(1.33x^3)\ cm^3$
9) $(79.51x^3)\ ft^3$
10) $(8x^3)\ mm^3$
11) $(0.027x^3)\ in^3$
12) $(238.33x^3)\ km^3$
13) 12
14) $21.66\ m^2$
15) $(138.24x^2)\ ft^2$
16) $(433.5x^2)\ mm^2$
17) $(0.96x^2)\ km^2$
18) $6x^2 - 12x + 6\ cm^2$

Rectangular Prism
1) $0.65\ m^3$
2) $3.19\ in^3$
3) $17.67\ m^3$
4) $(108x^3)\ cm^3$
5) $(18.4x^3)\ ft^3$
6) $(8.4x^3)\ m^3$
7) $(46x^2)\ cm^2$
8) $(2.54x^2)\ ft^2$
9) $(37x^2)\ in^2$
10) $(6x^2 - 2)\ m^2$

Cylinder
1) $663.86\ m^3$
2) $632.6\ cm^3$
3) $(8,488.55x^2)\ cm^3$
4) $(0.05x^3 + 0.05x^2)\ m^3$
5) $(6.104x^3)\ m^3$
6) $(0.5\ x^3 + 1.51x^2)\ in^3$
7) $(15.7x^2)\ m^2$
8) $(62.8x^2)\ cm^2$
9) $(5.97x^2)\ cm^2$
10) $(94.2x^2)\ m^2$

Pyramids and Cone
1) $2.35\ yd^3$
2) $31.68\ yd^3$
3) $26.59\ in^3$
4) $(18.84x^2)\ m^3$
5) $(619.10x)\ m^3$
6) $(0.262x + 0.262)\ cm^3$
7) $133.77\ yd^2$
8) $61.15\ m^2$
9) $(75.36x^2)\ in^2$
10) $(3.01)\ cm^2$

Chapter 9:
Statistics and Probability

Topics that you will practice in this chapter:

- ✓ Mean and Median
- ✓ Mode and Range
- ✓ Histograms
- ✓ Stem–and–Leaf Plot
- ✓ Pie Graph
- ✓ Probability Problems

Mathematics is no more computation than typing is literature.

– *John Allen Paulos*

Mean and Median

✎ Find Mean and Median of the Given Data.

1) 8, 9, 19, 3, 4

2) 11, 7, 35, 10, 17, 32, 24

3) 38, 9, 15, 17, 13

4) 50, 19, 2, 18, 6, 7

5) 25, 27, 13, 16, 6, 13, 54

6) 24, 364, 42, 57, 6, 68

7) 89, 98, 65, 45, 3, 4, 30, 42

8) 34, 15, 15, 17, 22, 29, 15

9) 2, 5, 10, 45, 8, 13, 35, 6

10) 20, 22, 18, 7, 2, 17, 44, 53

11) 33, 52, 81, 9, 45, 31

12) 19, 74, 51, 8, 12, 15, 9, 14

✎ Calculate.

13) In a javelin throw competition, five athletics score 45, 33, 53, 46 and 19 meters. What are their Mean and Median? _____

14) Eva went to shop and bought 5 apples, 9 peaches, 4 bananas, 7 pineapples and 8 melons. What are the Mean and Median of her purchase? _____

15) Bob has 19 black pen, 15 red pen, 27 green pens, 21 blue pens and one boxes of yellow pens. If the Mean and Median are 19 respectively, what is the number of yellow pens in box? _____

Mode and Range

✎ Find Mode and Rage of the Given Data.

1) 7, 4, 18, 9, 9, 3
   Mode: _____   Range: _____

2) 8, 8, 15, 14, 8, 5, 6, 18
   Mode: _____   Range: _____

3) 4, 4, 4, 15, 19, 24, 31, 5, 4
   Mode: _____   Range: _____

4) 10, 10, 9, 17, 14, 8, 20, 4
   Mode: _____   Range: _____

5) 5, 11, 3, 4, 3, 3
   Mode: _____   Range: _____

6) 13, 7, 7, 7, 7, 4, 12, 25, 8, 3
   Mode: _____   Range: _____

7) 1, 7, 9, 9, 24, 24, 24, 20, 34, 35
   Mode: _____   Range: _____

8) 9, 4, 7, 13, 13, 13, 9, 8, 15
   Mode: _____   Range: _____

9) 8, 8, 8, 5, 8, 7, 17, 16, 3, 9
   Mode: _____   Range: _____

10) 34, 34, 32, 14, 6, 14, 9, 14
    Mode: _____   Range: _____

11) 8, 8, 6, 8, 18, 10, 16, 15
    Mode: _____   Range: _____

12) 12, 12, 7, 11, 14, 12, 33, 5
    Mode: _____   Range: _____

✎ Calculate.

13) A stationery sold 21 pencils, 42 red pens, 25 blue pens, 26 notebooks, 21 erasers, 28 rulers and 27 color pencils. What are the Mode and Range for the stationery sells?

    Mode: _____   Range: _____

14) In an English test, eight students score 19, 10, 10, 17, 35, 35, 14 and 10. What are their Mode and Range? _____

15) What is the range of the first 6 odd numbers greater than 8?
    _____

Times Series

✎ **Use the following Graph to complete the table.**

Day	Distance (km)
1	
2	

The following table shows the number of births in the US from 2007 to 2012 (in millions).

Year	Number of births (in millions)
2007	4.25
2008	4.19
2009	4.55
2010	3.80
2011	3.25
2012	2.54

Draw a Time Series for the table.

Stem–and–Leaf Plot

✎ **Make stem ad leaf plots for the given data.**

1) 41, 44, 47, 40, 70, 45, 79, 77, 49, 44, 19, 10

2) 21, 87, 56, 20, 27, 23, 55, 82, 82, 53, 87, 58

3) 111, 47, 66, 44, 94, 117, 62, 114, 48, 112, 68, 99

4) 52, 25, 101, 58, 71, 26, 109, 53, 75, 29, 53, 108, 79

5) 51, 88, 9, 87, 81, 8, 3, 50, 85, 54, 9, 54, 5

6) 40, 93, 20, 25, 48, 92, 95, 52, 21, 44, 97, 29

Pie Graph

The circle graph below shows all Robert's expenses for last month. Robert spent $384 on his hobbies last month.

Answer following questions based on the Pie graph.

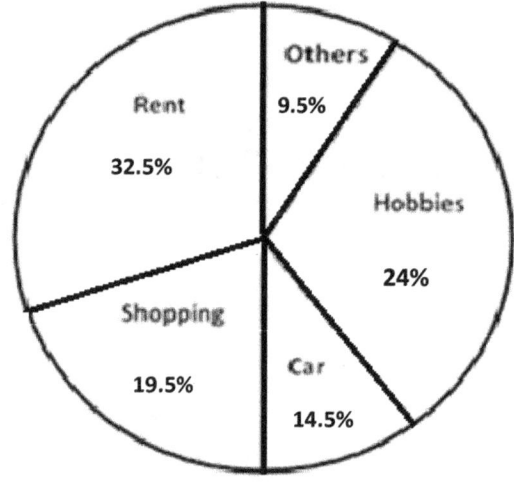

1) How much was Robert's total expenses last month? _____

2) How much did Robert spend on his car last month? _____

3) How much did Robert spend for shopping last month? _____

4) How much did Robert spend on his rent last month? _____

5) What fraction is Robert's expenses for his car and shopping out of his total expenses last month? _____

Probability Problems

✎ **Calculate.**

1) A number is chosen at random from 1 to 20. Find the probability of selecting number 8 or smaller numbers. _____

2) Bag A contains 16 red marbles and 6 green marbles. Bag B contains 12 black marbles and 18 orange marbles. What is the probability of selecting a green marble at random from bag A? What is the probability of selecting a black marble at random from Bag B? _____

3) A number is chosen at random from 1 to 25. What is the probability of selecting multiples of 5? _____

4) A card is chosen from a well-shuffled deck of 52 cards. What is the probability that the card will be a queen? _____

5) A number is chosen at random from 1 to 15. What is the probability of selecting a multiple of 4? _____

A spinner, numbered 1–8, is spun once. What is the probability of spinning …?

6) an Odd number? _____ 7) a multiple of 2? _____

8) a multiple of 5? _____ 9) number 10? _____

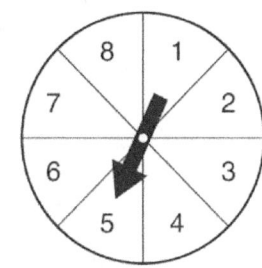

Answers of Worksheets – Chapter 9

Mean and Median

1) Mean: 8.6, Median: 8
2) Mean: 19.43, Median: 17
3) Mean: 18.4, Median: 15
4) Mean: 17, Median: 12.5
5) Mean: 22, Median: 16
6) Mean: 93.5, Median: 49.5
7) Mean: 47, Median: 43.5
8) Mean: 21, Median: 17
9) Mean: 15.5, Median: 9
10) Mean: 22.88, Median: 19
11) Mean: 41.83, Median: 39
12) Mean: 25.25, Median: 14.5
13) Mean: 39.2, Median: 45
14) Mean: 6.6, Median: 7
15) 13

Mode and Range

1) Mode: 9, Range: 15
2) Mode: 8, Range: 13
3) Mode: 4, Range: 27
4) Mode: 10, Range: 16
5) Mode: 3, Range: 8
6) Mode: 7, Range: 22
7) Mode: 24, Range: 34
8) Mode: 13, Range: 11
9) Mode: 8, Range: 14
10) Mode: 14, Range: 28
11) Mode: 8, Range: 12
12) Mode: 12, Range: 28
13) Mode: 21, Range: 21
14) Mode: 10, Range: 25
15) 10

Time series

Day	Distance (km)
1	356
2	352
3	285
4	540
5	365

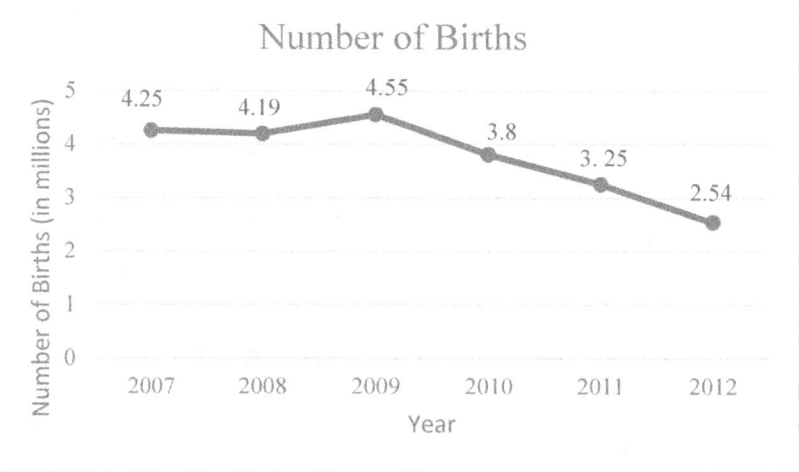

Stem–And–Leaf Plot

1)

Stem	leaf
1	0 9
4	0 1 4 4 5 7 9
7	0 7 9

2)

Stem	leaf
2	0 1 3 7
5	3 5 6 8
8	2 2 7 7

3)

Stem	leaf
4	4 7 8
6	2 6 8
9	4 9
11	1 2 4 7

4)

Stem	leaf
2	5 6 9
5	2 3 3 8
7	1 5 9
10	1 8 9

5)

Stem	leaf
0	3 5 8 9 9
5	0 1 4 4
8	1 5 7 8

6)

Stem	leaf
2	0 1 5 9
4	0 2 4 8
9	2 3 5 7

Pie Graph

1) $1,600

2) $232

3) $312

4) $520

5) $\frac{17}{50}$

Probability Problems

1) $\frac{2}{5}$

2) $\frac{3}{11}, \frac{2}{5}$

3) $\frac{1}{5}$

4) $\frac{1}{13}$

5) $\frac{1}{5}$

6) $\frac{1}{2}$

7) $\frac{1}{2}$

8) $\frac{1}{8}$

9) 0

Pre-Algebra Practice Tests

Time to Test

Time to refine your skill with a practice examination

Take a REAL Pre-Algebra test to simulate the test day experience. After you've finished, score your test using the answer key.

Before You Start

- You'll need a pencil, calculator, and a timer to take the test.
- It's okay to guess. You won't lose any points if you're wrong.
- After you've finished the test, review the answer key to see where you went wrong.

Graphing calculators are Not permitted for Pre-Algebra Tests

Good Luck!

Pre-Algebra Reference Materials

Circumference			
Circle	$C = 2\pi r$	or	$C = \pi d$

Area	
Triangle	$A = \frac{1}{2}bh$
Rectangle or Parallelogram	$A = bh$
Trapezoid	$A = \frac{1}{2}h(b_1 + b_2)$
Circle	$A = \pi r^2$

Surface Area	Lateral	Total
Prism	$S = ph$	$S = ph + 2B$
Cylinder	$S = 2\pi rh$	$S = 2\pi rh + 2\pi r^2$

Volume	
Prism or cylinder	$V = Bh$
Pyramid or Cone	$V = \frac{1}{3}Bh$
Sphere	$V = \frac{4}{3}\pi r^3$

Additional Information	
Pythagorean theorem	$a^2 + b^2 = c^2$
Simple interest	$I = prt$

Pre-Algebra Practice Test 1

1) The area of a circle is 49π. Which of the following can be the circumference of the circle?

   A. 14π C. 24π

   B. 18π D. 12π

2) You can buy 14 cans of green beans at a supermarket for $3.80. How much does it cost to buy 52 cans of green beans?

   A. $14.88   C. $14.11

   B. $24.56   D. $114.1

3) Which of the following is the solution of the following inequality?

   $$-9x + 25.5 > -12x + 14.5 + 8.5x$$

   A. $x < 2$ C. $x < -2$

   B. $x > 2$ D. $x > -2$

4) What is the perimeter of a square that has an area of 302.76 feet?

   Write your answer in the box below.

5) A tree 16 feet tall casts a shadow 34 feet long. Jack is 8 feet tall. How long is Jack's shadow?

   A. 19 ft   C. 17 ft

   B. 21 ft   D. 36 ft

Pre-Algebra Workbook

6) The price of a laptop is decreased by 35% to $580. What is its original price?

   A. 892.30

   B. 829.30

   C. 280.9

   D. 920.8

7) A container holds 4.8 gallons of water when it is $\frac{8}{41}$ full. How many gallons of water does the container hold when it's full?

   A. 24.6

   B. 34

   C. 14.45

   D. 8.42

8) If $(4^a)^b = 256$, then what is the value of $a \times b$?

   A. $4a$

   B. $5b$

   C. 4

   D. 5

9) A bag contains 20 balls: seven green, two black, two blue, eight red and one white. If 18 balls are removed from the bag at random, what is the probability that a white ball has been removed?

   A. $\frac{3}{20}$

   B. $\frac{18}{20}$

   C. $\frac{1}{20}$

   D. $\frac{1}{18}$

10) What is the x-intercept of the line with equation $-9x + 5y = 63$?

    A. -5

    B. -7

    C. $+9$

    D. $-\frac{9}{5}$

WWW.MathNotion.Com

11) Which of the following expressions is equivalent to

$8x^3y(3x + 2y)$?

A. $24yx^2 - 16x^3y$

B. $24x^2 - 16y^2$

C. $24yx^4 + 16x^3y^2$

D. $24x^3y^2 - 16yx^2$

12) A soccer team played 150 games and won 30 percent of them. How many games did the team win?

A. 110

B. 45

C. 95

D. 70

13) The equation of a line is given as: $y = -2x + 5$. Which of the following points does not lie on the line?

A. $(4, -3)$

B. $(1, 3)$

C. $(-1, 7)$

D. $(1, -1)$

14) The perimeter of the trapezoid below is 28 cm. What is its area?

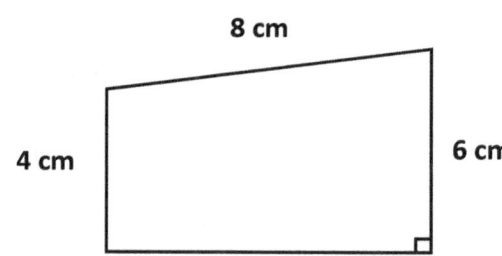

Write your answer in the box below.

15) Which graph does not represent y as a function of x?

A.
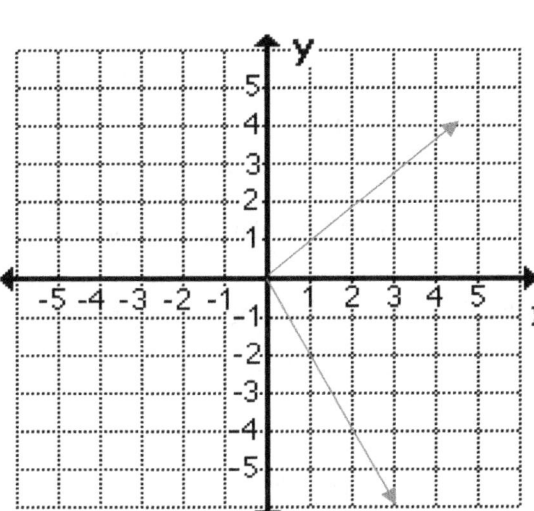

B.
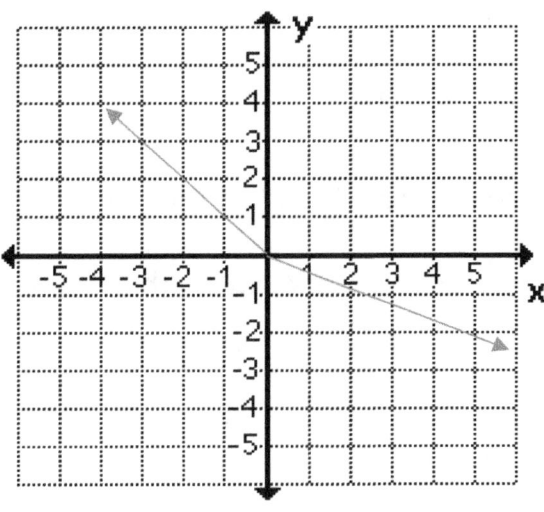

C.
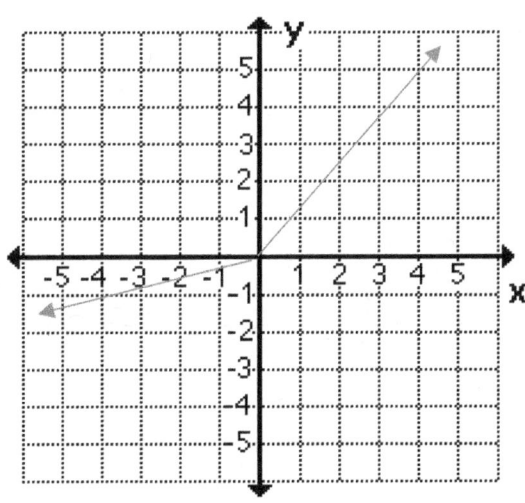

D.
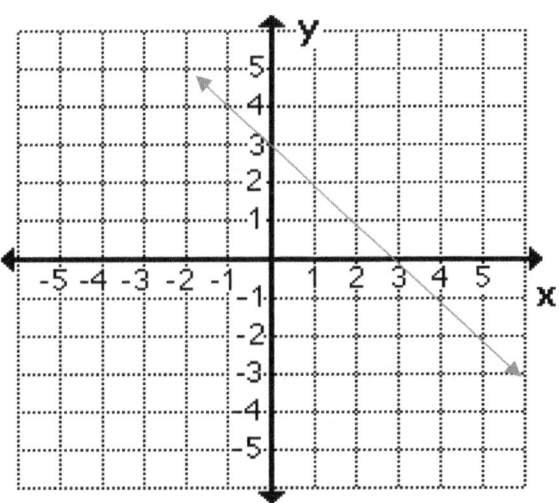

16) Which of the following is equivalent to $-43 < -7x + 6 < 6$?

A. $0 < x < 7$

B. $1 < x < 7$

C. $-7 < x < 0$

D. $-7 < x < 7$

17) A bank is offering 2.25% simple interest on a savings account. If you deposit $22,000, how much interest will you earn in five years?

A. $2,475

C. $2,780

B. $1,750

D. $4,840

18) Joe scored 34 out of 48 marks in Algebra, 26 out of 32 marks in science and 15 out of 22 marks in mathematics. In which subject his percentage of marks is best?

A. Mathematics

C. Algebra

B. Science

D. Mathematics and Science

19) What is the volume of the following triangular prism?

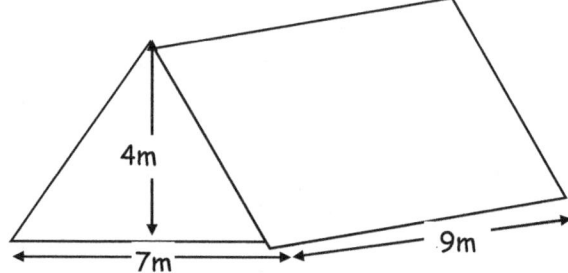

Write your answer in the box below.

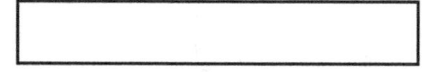

20) The marked price of a computer is E Euro. Its price decreased by 22% in March and later increased by 5 % in April. What is the final price of the computer in E Euro?

A. 0.819 E

C. 0.81 E

B. 0.098E

D. 0.918 E

Pre-Algebra Workbook

21) If $-3x + 7 = 2.5$, What is the value of $4x - \frac{3}{5}$?

   A. 4.5

   B. −4.5

   C. −5.4

   D. 5.4

22) Triangle ABC is graphed on a coordinate grid with vertices at A (4, 2), B (−3, 7) and C (1, 8). Triangle ABC is reflected over x axes to create triangle A' B' C'. Which order pair represents the coordinate of C'?

   A. (8, 1)

   B. (1, − 8)

   C. (−1, − 8)

   D. (−1, 8)

23) Which of the following point is the solution of the system of equations?

$$\begin{cases} 5x - 4y = 13 \\ 2x + 3y = -4 \end{cases}$$

   A. (−2, 1)

   B. (1, − 2)

   C. (−2, 2)

   D. (2, − 2)

24) A shirt costing $700 is discounted 30%. After a month, the shirt is discounted another 12%. Which of the following expressions can be used to find the selling price of the shirt?

   A. (700) (0.70) (0.12)

   B. (700) (0.30) (0.12)

   C. (700) (0.70) (0.88)

   D. (700) (0.30) (0.88)

25) What is the distance between the points (4, 2) and (− 8, − 3)?

   A. 13

   B. 11

   C. 14

   D. 15

Questions 26 and 27 are based on the following data

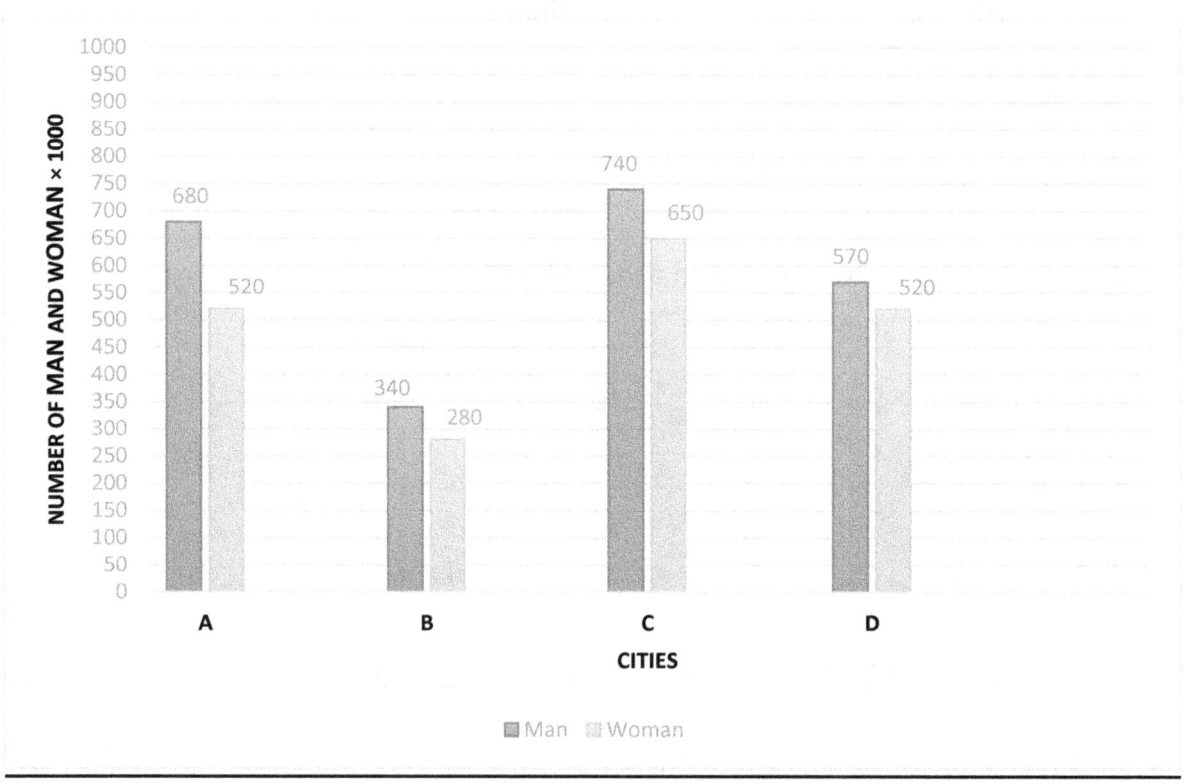

26) What's the ratio of percentage of men in city C to percentage of women in city B?

   A. 0.099   C. 1.18

   B. 0.809   D. 1.01

27) What's the maximum ratio of woman to man in the four cities?

    A. 0.90

    B. 0.91

    C. 0.87

    D. 0.97

28) Line m passes through the point $(4, -7)$. Which of the following **CANNOT** be the equation of line m?

    A. $y = -7$

    B. $y = -2x + 1$

    C. $y = -2 + x$

    D. $y = -3x + 5$

29) The sum of five numbers is 98. If another number is added to these five numbers, the average of the six numbers is 32. What is the sixth number?

    A. 102

    B. 92

    C. 88

    D. 94

30) David owed $15,435. After making 38 payments of $324 each, how much did he have left to pay?

    A. $3,382

    B. $4,231

    C. $3,123

    D. $4,301

31) Simplify the expression.

$$(5x^3 - 7x^2 - 3x^4) - (5x^2 - 6x^4 - x^3)$$

    A. $-3(x^4 + 3x^3 - 4x^2)$

    B. $3(2x^4 + 2x^3 - 4x^2)$

    C. $3(x^4 + 2x^3 - 2x^2)$

    D. $3(x^4 + 2x^3 - 4x^2)$

Pre-Algebra Workbook

32) The average of six consecutive numbers is 22.5. What is the smallest number?

   A. 20

   B. 19

   C. 23

   D. 21

33) The price of a laptop is decreased by 16% to $378. What is its original price?

   A. 450

   B. 410.5

   C. 840

   D. 240

34) The average weight of 16 girls in a class is 40 kg and the average weight of 29 boys in the same class is 50 kg. What is the average weight of all the 45 students in that class?

   A. 40.56 Kg

   B. 42.56 Kg

   C. 46.88 Kg

   D. 46.44 Kg

35) Liam's average (arithmetic mean) on four mathematics tests is 76. What should Liam's score be on the next test to have an overall of 72 for all the tests?

   A. 56

   B. 58

   C. 68

   D. 70

36) Which of the following equations has a graph that is a straight line?

   A. $y = 2x^2 - 7$

   B. $5y^2 - 3x^2 = 7$

   C. $x - y = 8$

   D. $3x + xy = 8$

37) An angle is equal to one seventh of its supplement. What is the measure of that angle?

   A. 26

   B. 18

   C. 22.5

   D. 24.5

38) Two ninth of 27 is equal to $\frac{6}{11}$ of what number?

   A. 36

   B. 22

   C. 44

   D. 11

39) A card is drawn at random from a standard 52–card deck, what is the probability that the card is of Hearts or diamonds? (The deck includes 13 of each suit clubs, diamonds, hearts, and spades)

   A. $\frac{1}{2}$

   B. $\frac{2}{13}$

   C. $\frac{2}{52}$

   D. $\frac{3}{13}$

40) The score of Zoe was one sixth of Emma and the score of Harper was twice that of Emma. If the score of Harper was 84, what is the score of Zoe?

   A. 14

   B. 7

   C. 16

   D. 9

Pre-Algebra Practice Test 2

Pre-Algebra Workbook

1) Right triangle ABC has two legs of lengths 18 cm (AB) and 24 cm (AC). What is the length of the third side (BC)?

   A. 15 cm

   B. 28 cm

   C. 42 cm

   D. 30 cm

2) When a number is subtracted from 80 and the difference is divided by that number, the result is 4. What is the value of the number?

   A. 16

   B. 20

   C. 18

   D. 12

3) A bank is offering 2.4% simple interest on a savings account. If you deposit $10,000, how much interest will you earn in three years?

   A. $140

   B. $1,540

   C. $360

   D. $720

4) In a party, 15 soft drinks are required for every 9 guests. If there are 162 guests, how many soft drinks is required?

   A. 135

   B. 270

   C. 180

   D. 320

5) A chemical solution contains 16% alcohol. If there is 9.2 ml of alcohol, what is the volume of the solution?

   A. 57.5ml

   B. 435 ml

   C. 575 ml

   D. 685 ml

6) What is the area of the shaded region?

   A. 72

   B. 112

   C. 40

   D. 86

7) A rope weighs 210 grams per meter of length. What is the weight in kilograms of 22.3 meters of this rope? (1 kilograms = 1000 grams)

   A. 4,683

   B. 0.4683

   C. 46.83

   D. 4.683

8) The ratio of boys to girls in a school is 5: 9. If there are 364 students in a school, how many boys are in the school.

   Write your answer in the box below.

9) In two successive years, the population of a town is increased by 8% and 14%. What percent of the population is increased after two years?

   A. 26.1%

   B. 23.12%

   C. 123.12%

   D. 45.6 %

10) Which graph shows a non–proportional linear relationship between x and y?

A.   B.

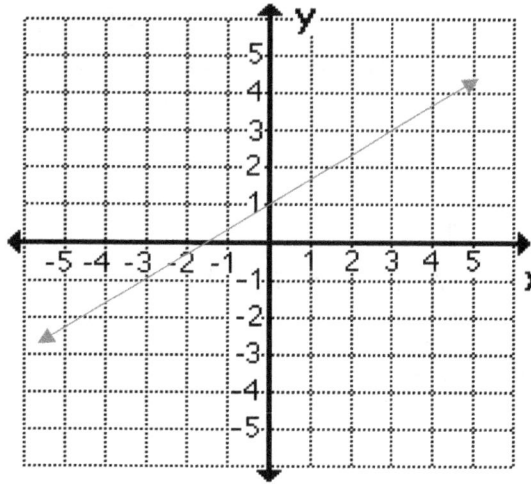

C.   D.

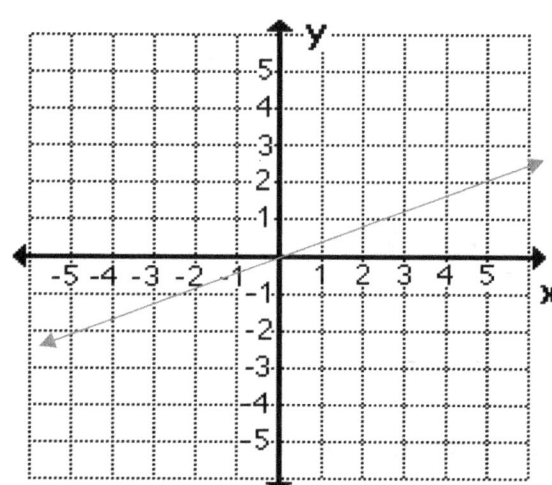
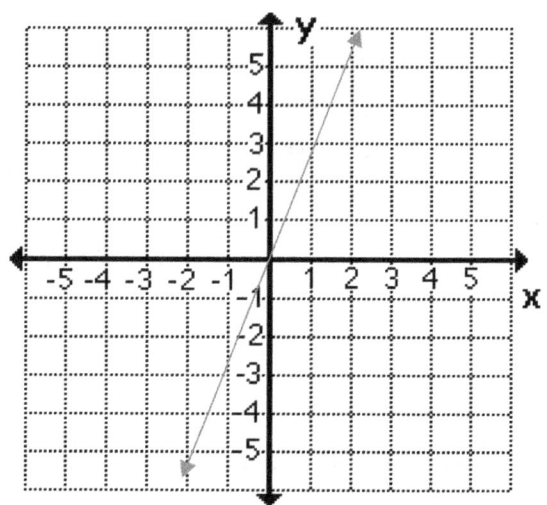

11) Four years ago, Amy was three times as old as Mike was. If Mike is 12 years old now, how old is Amy?

A. 18   C. 36

B. 28   D. 24

12) What is the value of $|-44+15| - |6(-3)|$?

   A. 11

   B. −11

   C. −13

   D. +13

13) What is the solution of the following system of equations?

$$\begin{cases} \dfrac{x}{6}+\dfrac{y}{4}=2 \\ \dfrac{-5y}{6}-2x=-11 \end{cases}$$

   A. $x=3, y=6$

   B. $x=2, y=8$

   C. $x=4, y=8$

   D. $x=-6, y=8$

14) If a gas tank can hold 45 gallons, how many gallons does it contain when it is $\frac{2}{9}$ full?

   A. 42

   B. 30

   C. 10

   D. 20

15) What is the length of BC in the following figure if AB = 3, DF = 4 and BD = 42?

   A. 14

   B. 18

   C. 9

   D. 36

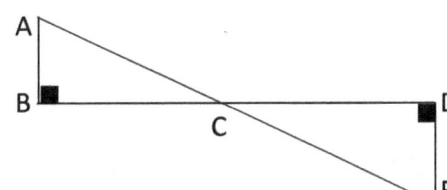

16) What is the product of all possible values of x in the following equation?

$$|-8x + 4| = 28$$

    A. -3

    B. -12

    C. 12

    D. 4

17) In the rectangle below if $y > 9$ cm and the area of rectangle is 90 cm² and the perimeter of the rectangle is 38 cm, what is the value of x and y respectively?

    A. 9, 10

    B. 7, 12

    C. 12, 7

    D. 5, 18

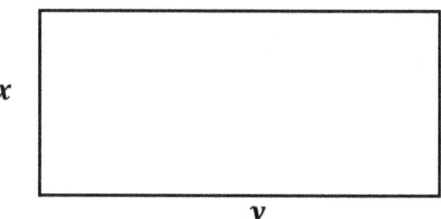

18) A football team had $26,000 to spend on supplies. The team spent $18,000 on new balls. New sport shoes cost $102 each. Which of the following inequalities represent the number of new shoes the team can purchase?

    A. $102x + 18,000 \leq 26,000$

    B. $102x + 18,000 \geq 26,000$

    C. $18,000 + 102x \geq 26,000$

    D. $18,000 + 102x \leq 26,000$

19) A $70 shirt now selling for $49 is discounted by what percent?

   A. 55 %

   B. 40 %

   C. 30 %

   D. 35 %

20) How much interest is earned on a principal of $11,000 invested at an interest rate of 2.5% for six years?

   A. $1,650

   B. $2,400

   C. $1,880

   D. $16,560

21) The price of a car was $56,000 in 2014, $42,000 in 2015 and $31,500 in 2016. What is the rate of depreciation of the price of car per year?

   A. 25 %

   B. 10 %

   C. 35 %

   D. 55 %

22) The Jackson Library is ordering some bookshelves. If x is the number of bookshelves the library wants to order, which each cost $86 and there is a one-time delivery charge of $670, which of the following represents the total cost, in dollar, per bookshelf?

   A. $86x + 670$

   B. $86 + 670x$

   C. $\dfrac{86x+670}{86}$

   D. $\dfrac{86x+670}{x}$

23) The following table represents the value of x and function $f(x)$. Which of the following could be the equation of the function $f(x)$?

A. $f(x) = 2x^2 + 6$

B. $f(x) = 4x^2 - 6x + 6$

C. $f(x) = \sqrt{2x + 6}$

D. $f(x) = 2\sqrt{2x} + 6$

x	$f(x)$
0	6
2	10
8	14
32	22

24) The circle graph below shows all Mr. Wilson's expenses for last month. If he spent $880 on his car, how much did he spend for his rent?

A. $1,720

B. $1,360

C. $4,000

D. 2,620

Mr. Wilson's Monthly Expenses

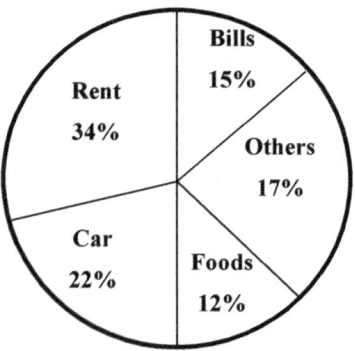

25) The sum of five different negative integers is -50. If the smallest of these integers is -12, what is the largest possible value of one of the other five integers?

A. -8

B. -4

C. -6

D. -12

26) In the following figure, point M lies on the line A, what is the value of y if $x = 10$?

A. 16

B. 38

C. 19

D. 40

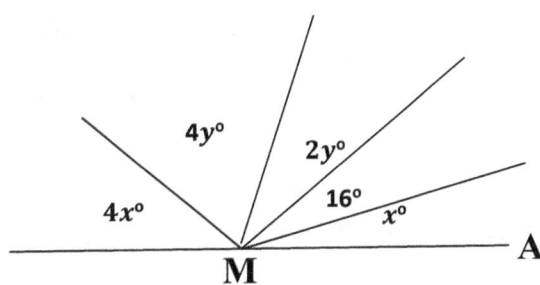

27) What is the smallest integer whose square root is greater than 7?

A. 121

B. 52

C. 81

D. 58

28) What is the area of the shaded region if the diameter of the bigger circle is 10 inches and the diameter of the smaller circle is 6 inches?

A. 30 π

B. 24 π

C. 16π

D. 12 π

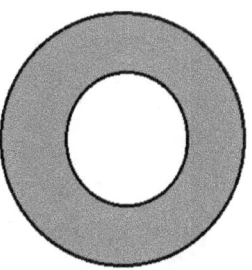

29) What is the sum of $\sqrt{3x+6}$ and $\sqrt{x} - 4$ when $\sqrt{x} = -5$?

A. −9

B. −6

C. 8

D. 10

Pre-Algebra Workbook

30) What is the area of an isosceles right triangle that has one leg that measures 14 cm?

Write your answer in the box below.

☐

31) If x is directly proportional to the square of y, and $y = 3$ when $x = 63$, then if $x = 175$ $y = ?$

A. 5

B. 10

C. 12

D. 25

32) What is the value of x in the following figure?

A. 50

B. 70

C. 60

D. 100

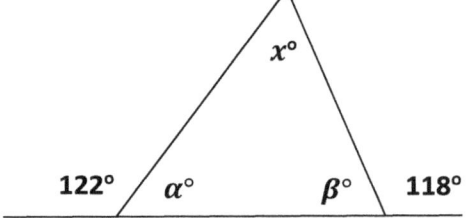

33) A swimming pool holds 1,344 cubic feet of water. The swimming pool is 32 feet long and 14 feet wide. How deep is the swimming pool?

Write your answer in the box below.

☐

34) Jack earns $500 for his first 25 hours of work in a week and is then paid 1.5 times his regular hourly rate for any additional hours. This week, Jack needs $920 to pay his rent, bills and other expenses. How many hours must he work to make enough money in this week?

A. 46

B. 48

C. 34

D. 36

Questions 35, 36 and 37 are based on the following data

Types of air pollutions in 10 cities of a country

Type of Pollution	Number of Cities									
A	■	■	■	■						
B	■	■								
C	■									
D	■	■	■	■	■	■	■	■	■	
E	■	■	■							
	1	2	3	4	5	6	7	8	9	10

35) If a is the mean (average) of the number of cities in each pollution type category, b is the mode, and c is the median of the number of cities in each pollution type category, then which of the following must be true?

A. $a = b < c$

B. $b = a = c$

C. $c < a$

D. $b < c < a$

36) How many cities should be added to type of pollutions B until the ratio of cities in type of pollution B to cities in type of pollution D will be 1?

A. 3

B. 6

C. 9

D. 12

37) What percent of cities are in the type of pollution A, C, and E respectively?

A. 60%, 20%, 40%

B. 1.60%, 1.20%, 1.40%

C. 1.40%, 1.60%, 1.70%

D. 20%, 50%, 80%

38) In the following right triangle, if the sides AB and AC become half longer, what will be the ratio of the perimeter of the triangle to its area?

A. $\frac{2}{3}$

B. $\frac{1}{2}$

C. 2

D. 1

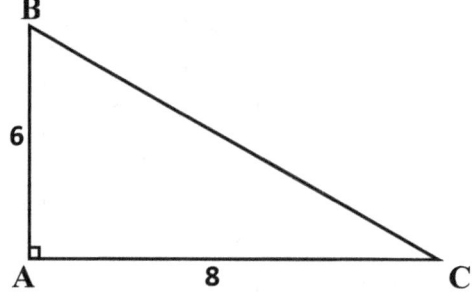

39) The capacity of a red box is 30% bigger than the capacity of a blue box. If the red box can hold 78 equal sized books, how many of the same books can the blue box hold?

A. 25

B. 60

C. 40

D. 50

40) A taxi driver earns $15 per hour work. If he works 6 hours a day, and he uses 3-liters Petrol in 2 hours with price $2.2 for 1-liter. How much money does he earn in one day?

    A. $90

    B. $82.5

    C. $19.8

    D. $70.2

Answers and Explanations
Pre-Algebra Practice Test

Answer Key

✻ Now, it's time to review your results to see where you went wrong and what areas you need to improve!

Pre-Algebra Practice Tests

Practice Test - 1						Practice Test - 2					
1	A	16	B	31	D	1	D	16	B	31	A
2	C	17	A	32	A	2	A	17	A	32	C
3	A	18	B	33	A	3	D	18	A	33	3
4	69.6	19	126	34	D	4	B	19	C	34	C
5	C	20	A	35	A	5	C	20	A	35	C
6	A	21	D	36	C	6	A	21	A	36	B
7	A	22	B	37	C	7	D	22	D	37	A
8	C	23	B	38	D	8	130	23	D	38	C
9	C	24	C	39	A	9	B	24	B	39	C
10	B	25	A	40	B	10	A	25	A	40	D
11	C	26	C			11	B	26	C		
12	B	27	B			12	A	27	B		
13	D	28	C			13	A	28	C		
14	50	29	D			14	C	29	D		
15	A	30	C			15	B	30	98		

Practice Test 1
Pre-Algebra Explanations

1) Answer: A

Use the formula for area of circles.

Area = $\pi r^2 \Rightarrow 49\pi = \pi r^2 \Rightarrow 49 = r^2 \Rightarrow r = 7$

Radius of the circle is 7. Now, use the circumference formula:

Circumference = $2\pi r = 2\pi(7) = 14\pi$

2) Answer: C

Let x be the number of cans. Write the proportion and solve for x.

$\frac{14 \text{ cans}}{\$3.80} = \frac{52 \text{ cans}}{x} \Rightarrow x = \frac{3.80 \times 52}{14} \Rightarrow x = \$14.11$

3) Answer: A

$-9x + 25.5 > -12x + 14.5 + 8.5x \rightarrow$ Combine like terms:

$-9x + 25.5 > -3.5x + 14.5$ Subtract $9x$ from both sides: $25.5 > 5.5x + 14.5$

Add -14.5 both sides of the inequality.

$11 > 5.5x$, Divide both sides by 5.5. $\Rightarrow \frac{11}{5.5} > x \rightarrow x < 2$

4) Answer: 69.6 feet.

Area of a square: S = a2 $\Rightarrow 302.76 = a^2 \Rightarrow$ a = 17.4

Perimeter of a square: P = 4a $\Rightarrow$ P = 4 × 17.4 $\Rightarrow$ P = 69.6

5) Answer: C

Write the proportion and solve for the missing number.

$\frac{16}{34} = \frac{8}{x} \rightarrow 16x = 8 \times 34 = 272 \rightarrow 16x = 272 \rightarrow x = \frac{272}{16} = 17$

6) Answer: A

Let x be the original price.

If the price of a laptop is decreased by 35% to $580, then:

$65\% \text{ of } x = 580 \Rightarrow 0.65x = 580 \Rightarrow x = 580 \div 0.65 = 892.30$

7) Answer: A

let x be the number of gallons of water the container holds when it is full.

Then; $\frac{8}{41}x = 4.8 \rightarrow x = \frac{41 \times 4.8}{8} = 24.6$

8) Answer: C

$(4^a)^b = 256 \rightarrow 4^{ab} = 256 \rightarrow 256 = 4^4 \rightarrow 4^{ab} = 4^4 \rightarrow ab = 4$

9) Answer: C

If 18 balls are removed from the bag at random, there will be one ball in the bag.

The probability of choosing a white ball is 1 out of 20. Therefore, the probability of not choosing a white ball is 18 out of 20 and the probability of having not a white ball after removing 18 balls is the same.

10) Answer: B

The value of y in the x-intercept of a line is zero. Then:

$y = 0 \rightarrow -9x + 5(0) = 63 \rightarrow -9x = 63 \rightarrow x = \frac{-63}{9} = -7$

then, x-intercept of the line is -7

11) Answer: C

Use distributive property:

$8x^3y(3x + 2y) = 8x^3y(3x) + 8x^3y(2y) = 24yx^4 + 16x^3y^2$

12) Answer: B

$150 \times \frac{30}{100} = 45.$

13) Answer: D

$y = -2x + 5$

$(4, -3) \Rightarrow -3 = -2(4) + 5 \Rightarrow -3 = -3$

$(1, 3) \Rightarrow 3 = -2(1) + 5 \Rightarrow 3 = 3$

$(-1, 7) \Rightarrow 7 = -2(-1) + 5 \Rightarrow 7 = 7$

$(1, -1) \Rightarrow -1 = -2(1) + 5 \Rightarrow -1 \neq 3$

14) Answer: 50.

The perimeter of the trapezoid is 28 cm.

Therefore, the missing side (height) is $= 28 - 8 - 6 - 4 = 10$

Area of a trapezoid: A = $\frac{1}{2}$ h (b$_1$ + b$_2$) = $\frac{1}{2}$ (10) (4 + 6) = 50

15) Answer: A

A graph represents y as a function of x if $x_1 = x_2 \rightarrow y_1 = y_2$

In choice A, for each x, we have two different values for y.

16) Answer: B

$-43 < -7x + 6 < 6 \rightarrow$ Subtract 6 to all sides.

$-43 - 6 < -7x + 6 - 6 < 6 - 6$

$\rightarrow -49 < -7x < 0 \rightarrow$ Divide all sides by -7. (Remember that when you divide all sides of an inequality by a negative number, the inequality sing will be swapped. < becomes >)

$\frac{-49}{-7} < \frac{-7x}{-7} < \frac{0}{-7} \Rightarrow 7 > x > 0$, or $0 < x < 7$

17) Answer: A

Use simple interest formula: $I = prt$ (I = interest, p = principal, r = rate, t = time)

$I = (22,000)(0.0225)(5) = 2,475$

18) Answer: B

Compare each mark:

In Algebra Joe scored 34 out of 48 in Algebra. It means Joe scored 70.83% of the total mark. $\frac{34}{48} = \frac{x}{100} \Rightarrow x = 70.83\%$

Joe scored 26 out of 32 in science. It means Joe scored 81.25% of the total mark. $\frac{26}{32} = \frac{x}{100}$

$\Rightarrow x = 81.25\%$

Joe scored 15 out of 22 in mathematic that it means 68.18% of total mark.

$\frac{15}{22} = \frac{x}{100} \Rightarrow x = 68.18\%$

Therefore, his score in Science is higher than his other scores.

19) Answer: 126 m³.

Use the volume of the triangular prism formula. V = $\frac{1}{2}$ (length) (base) (high)

V = $\frac{1}{2} \times 9 \times 7 \times 4 \Rightarrow$ V = 126 m³

20) Answer: A

To find the discount, multiply the number by (100% − rate of discount).

Therefore, for the first discount we get: $(100\% - 22\%)(E) = (0.78)E$

For increase of 5 %:

$(0.78)E \times (100\% + 5\%) = (0.78)(1.05) = 0.819E$

21) Answer: D

$-3x + 7 = 2.5 \rightarrow -3x = 2.5 - 7 = -4.5 \rightarrow x = \frac{-4.5}{-3} = 1.5$

Then; $4x - \frac{3}{5} = 4(1.5) - \frac{3}{5} = 6 - 0.6 = 5.4$.

22) Answer: B

When a point is reflected over x axes, the (y) coordinate of that point changes to $(-y)$ while its x coordinate remains the same.

B $(1, 8) \rightarrow$ B' $(1, -8)$

23) Answer: B

Solving Systems of Equations by Elimination

$\begin{array}{l} 5x - 4y = 13 \\ 2x + 3y = -4 \end{array}$ Multiply the first equation by -2, and second equation by 5, then add two equations.

$\begin{array}{l} -2(5x - 4y = 13) \\ 5(2x + 3y = -4) \end{array} \Rightarrow \begin{array}{l} -10x + 8y = -26 \\ 10x + 15y = -20 \end{array} \Rightarrow 23y = -46 \Rightarrow y = -2$.

$2x + 3y = -4, 2x + 3(-2) = -4$, then: $x = 1, (1, -2)$

24) Answer: C

To find the discount, multiply the number by (100% − rate of discount).

Therefore, for the first discount we get: $(700)(100\% - 30\%) = (400)(0.70)$

For the next 12 % discount: $(700)(0.70)(0.88)$.

25) Answer: A

Use distance formula:

$C = \sqrt{(x_A - x_B)^2 + (y - y_B)^2} \Rightarrow C = \sqrt{(4 - (-8))^2 + (2 - (-3))^2}$

$C = \sqrt{(12)^2 + (5)^2} \Rightarrow C = \sqrt{144 + 25} \Rightarrow C = \sqrt{169} = 13$

26) Answer: C

Percentage of women in city C = $\frac{740}{1,390} \times 100 = 53.23\%$

Percentage of men in city B = $\frac{280}{620} \times 100 = 45.16\%$

Percentage of women in city C to percentage of men in city B: $\frac{53.23}{45.16} = 1.18$

27) Answer: B

Ratio of women to men in city A: $\frac{520}{680} = 0.76$

Ratio of women to men in city B: $\frac{280}{340} = 0.82$

Ratio of women to men in city C: $\frac{650}{740} = 0.87$

Ratio of women to men in city D: $\frac{520}{570} = 0.91$

28) Answer: C

Solve for each equation: $(4, -7)$

$y = -7 \Rightarrow -7 = -7$

$y = -2x + 1 \Rightarrow -7 = -2(4) + 1 \Rightarrow -7 = -7$

$y = -2 + x \Rightarrow -7 = -2 + 4 \Rightarrow 2 \neq -3$

$y = -3x + 5 \Rightarrow -7 = -3(4) + 5 \Rightarrow -7 = -7$

29) Answer: D

$a + b + c + d + e = 98 \Rightarrow \frac{a+b+c+d+e+f}{6} = 32 \Rightarrow a + b + c + d + e + f = 192$

$\Rightarrow 98 + f = 192 \Rightarrow f = 192 - 98 = 94$

30) Answer: C

$38 \times \$324 = \$12,312$ Payable amount is: $\$15,435 - 12,312 = 3,123$

31) Answer: D

Simplify and combine like terms.

$(5x^3 - 7x^2 - 3x^4) - (5x^2 - 6x^4 - x^3) \Rightarrow (5x^3 - 7x^2 - 3x^4) - 5x^2 + 6x^4 + x^3$

$\Rightarrow 3x^4 + 6x^3 - 12x^2 = 3(x^4 + 2x^3 - 4x^2)$.

32) Answer: A

Let x be the smallest number. Then, these are the numbers:

$x, x+1, x+2, x+3, x+4, x+5$

$$\text{average} = \frac{\text{sum of terms}}{\text{number of terms}} \Rightarrow 22.5 = \frac{x+(x+1)+(x+2)+(x+3)+(x+4)+(x+5)}{6} \Rightarrow 22.5 = \frac{6x+15}{6}$$

$\Rightarrow 135 = 6x + 15 \Rightarrow 120 = 6x \Rightarrow x = 20$

33) Answer: A

Let x be the original price.

If the price of a laptop is decreased by 16% to $378, then:

$84\%\ of\ x = 378 \Rightarrow 0.84\ x = 378 \Rightarrow x = 378 \div 0.84 = 450$

34) Answer: D

The sum of the weight of all girls is: $16 \times 40 = 640$ kg

The sum of the weight of all boys is: $29 \times 50 = 1{,}450$ kg

The sum of the weight of all students is: $640 + 1{,}450 = 2{,}090$ kg

$\text{average} = \frac{\text{sum of terms}}{\text{number of terms}}$; $\text{average} = \frac{2{,}090}{45} = 46.44$

35) Answer: A

$\frac{a+b+c+d}{4} = 76 \Rightarrow a+b+c+d = 304$

$\frac{a+b+c+d+e}{5} = 72 \Rightarrow a+b+c+d+e = 360$

$304 + e = 360 \Rightarrow e = 360 - 304 = 56$

36) Answer: C

$x - y = 8$ has a graph that is a straight line.

All other options are not equations of straight lines.

37) Answer: C

The sum of supplement angles is 180. Let x be that angle. Therefore,

$x + 7x = 180;\ 8x = 180$, divide both sides by 8: $x = 22.5$

38) Answer: D

Let x be the number. Write the equation and solve for x.

$\frac{2}{9} \times 27 = \frac{6}{11}x \Rightarrow \frac{2 \times 27}{9} = \frac{6x}{11}$, use cross multiplication to solve for x.

$22 \times 27 = 6x \times 9 \Rightarrow 594 = 54x \Rightarrow x = 11$

39) Answer: A

The probability of choosing a Hearts is $\frac{13}{52} + \frac{13}{52} = \frac{1}{2}$

40) Answer: B

If the score of Harper was 84, therefore the score of Emma is 42. Since, the score of Zoe was one sixth of Emma, therefore, the score of Zoe is 7.

Practice Test 2
Pre-Algebra Explanations

1) Answer: D

Use Pythagorean Theorem: $a^2 + b^2 = c^2$

$18^2 + 24^2 = C^2 \Rightarrow 324 + 576 = C2 \Rightarrow 900 = c^2 \Rightarrow c = 30$

2) Answer: A

Let x be the number. Write the equation and solve for x. $\frac{(80-x)}{x} = 4$ (cross multiply)

$(80 - x) = 4x$, then add x both sides. $80 = 5x$, now divide both sides by 5. $\Rightarrow x = 16$.

3) Answer: D

Use simple interest formula: $I = prt$; (I=interest. p=principal. r=rate. t=time)

$I = (10,000)(0.024)(3) = 720$

4) Answer: B

Let x be the number of soft drinks for 132 guests. Write the proportion and solve for x.

$\frac{15 \text{ soft drinks}}{9 \text{ guests}} = \frac{x}{162 \text{ guests}} \Rightarrow x = \frac{162 \times 15}{9} \Rightarrow x = 270$

5) Answer: C

16% of the volume of the solution is alcohol. Let x be the volume of the solution. Then:

16% of x = 9.2 ml

$0.16\ x = 9.2 \Rightarrow \frac{16x}{100} = \frac{92}{10}$ cross multiply; $16x = 9,200 \Rightarrow$ (devide by 16) $x = 575$

6) Answer: A

Use the area of rectangle formula (S = a × b).

To find area of the shaded region subtract smaller rectangle from bigger rectangle.

$S_1 - S_2 = (7\ ft \times 16ft) - (4ft \times 10ft) \Rightarrow S_1 - S_2 = 72ft$.

7) Answer: D

The weight of 22.3 meters of this rope is: $22.3 \times 210g = 4,683g$

1 kg = 1,000 g, therefore, 4,683 g ÷ 1000 = 4.683kg

8) Answer: 130

The ratio of boy to girls is 5:9. Therefore, there are 5 boys out of 14 students. To find the answer, first divide the total number of students by 14, then multiply the result by 5.

$364 \div 14 = 26 \Rightarrow 26 \times 5 = 130$

9) Answer: B

the population is increased by 8% and 14%. 8% increase changes the population to 108% of original population.

For the second increase, multiply the result by 114%.

$(1.08) \times (1.14) = (1.2312) = 123.12\%$

23.12 percent of the population is increased after two years.

10) Answer: A

A linear equation is a relationship between two variables, x and y, that can be put in the form $y = mx + b$.

A non-proportional linear relationship takes on the form $y = mx + b$, where $b \neq 0$ and its graph is a line that does not cross through the origin.

11) Answer: B

Four years ago, Amy was three times as old as Mike. Mike is 12 years now. Therefore, 4 years ago Mike was 8 years.

Four years ago, Amy was: $A = 8 \times 3 = 24$

Now Amy is 28 years old: $24 + 4 = 28$

12) Answer: A

$|-44 + 15| - |6(-3)| = |-29| - |-18| = 29 - 18 = 11$

13) Answer: A

$\begin{cases} \frac{x}{6} + \frac{y}{4} = 2 \\ \frac{-5y}{6} - 2x = -11 \end{cases}$ → Multiply the top equation by 12. Then,

$\begin{cases} 2x + 3y = 24 \\ \frac{-5y}{6} - 2x = -11 \end{cases}$ → Add two equations.

$\frac{13}{6} y = 13 \rightarrow y = 6$, plug in the value of y into the first equation → $x = 3$

14) Answer: C

$$\frac{2}{9} \times 45 = \frac{90}{9} = 10$$

15) Answer: B

Two triangles ΔABC and ΔCD are similar. Then:

$$\frac{AB}{DF} = \frac{BC}{CD} \to \frac{3}{4} = \frac{x}{42-x} \to 126 - 3x = 4x \to 7x = 126 \to x = 18$$

16) Answer: B

To solve absolute values equations, write two equations.

$-8x + 4$ can equal positive 28, or negative 28. Therefore,

$-8x + 4 = 28 \Rightarrow -8x = 24 \Rightarrow x = -3$

$-8x + 4 = -28 \Rightarrow -8x = -28 - 4 = -32 \Rightarrow x = 4$

17) Answer: A

The perimeter of the rectangle is: $2x + 2y = 38$

$\to x + y = 19 \to x = 19 - y$

The area of the rectangle is: $x \times y = 90 \to (19 - y)(y) = 90$

$\to y^2 - 19y + 90 = 0$

Solve the quadratic equation by factoring method.

$(y - 9)(y - 10) = 0 \to y = 9$ (Unacceptable, because y must be greater than 9) or $y = 10$; If $y = 10 \to x = 19 - y \to x = 19 - 10 \to x = 9$

18) Answer: A

Let x be the number of new shoes the team can purchase. Therefore, the team can purchase $102\ x$.

The team had $26,000 and spent $18,000. Now the team can spend on new shoes $8,000 at most. Now, write the inequality: $102x + 18,000 \leq 26,000$

19) Answer: C

Use the formula for Percent of Change

$$\frac{\text{New Value} - \text{Old Value}}{\text{Old Value}} \times 100\ \% \Rightarrow \frac{49 - 70}{70} \times 100\ \% = -30\ \%$$

(negative sign here means that the new price is less than old price).

Pre-Algebra Workbook

20) Answer: A

Use simple interest formula: $I = prt$ (I = interest, p = principal, r = rate, t = time)

$I = (11,000)(0.025)(6) = 1,650$

21) Answer: A

Use this formula: Percent of Change $= \frac{New\ Value - Old\ Value}{Old\ Value} \times 100\ \%$

$\frac{42,000-56,000}{56,000} \times 100\ \% = -25\ \%$ and $\frac{31,500-42,000}{42,000} \times 100\% = -25\ \%$

22) Answer: D

The amount of money for x bookshelf is: $86x$

Then, the total cost of all bookshelves is equal to: $86x + 670$

The total cost, in dollar, per bookshelf is: $\frac{Total\ cost}{number\ of\ items} = \frac{86x+670}{x}$

23) Answer: D

A. $f(x) = 2x^2 + 6$ if $x = 8 \to f(8) = 2(8)^2 + 6 = 134 \neq 14$

B. $f(x) = 4x^2 - 6x + 6$ if $x = 8 \to f(8) = 4(8)^2 - 6 \times 8 + 6 = 214 \neq 14$

C. $f(x) = \sqrt{2x+6}$ if $x = 8 \to f(8) = \sqrt{2(8)+6} = \sqrt{22} \neq 14$

D. $f(x) = 2\sqrt{2x} + 6$ if $x = 8 \to f(8) = 2\sqrt{2(8)} + 6 = 14 = 14$

24) Answer: B

Let x be all expenses, then $\frac{22}{100}x = \$880 \to x = \frac{100 \times \$880}{22} = \$4,000$

He spent for his rent: $\frac{34}{100} \times \$4,000 = \$1,360$

25) Answer: A

The smallest number is -12. To find the largest possible value of one of the other five integers, we need to choose the smallest possible integers for three of them. Let x be the largest number. Then: $-50 = (-12) + (-11) + (-10) + (-9) + x$

$\to -50 = -42 + x \to x = -50 + 42 = -8$

26) Answer: C

The angles on a straight line add up to 180 degrees. Then:

$x + 16 + 2y + 4y + 4x = 180$

WWW.MathNotion.Com

Then, $5x + 6y = 180 - 16 \to 5(10) + 6y = 164 \to 6y = 164 - 50 = 114 \to y = 19$

27) **Answer: B**

Square root of 121 is $\sqrt{121} = 11 > \sqrt{49} = 7$

Square root of 52 is $\sqrt{52} = \sqrt{49 + 3} > \sqrt{49} = 7$

Square root of 81 is $\sqrt{81} = 9 > 7$

Square root of 58 is $\sqrt{58} = \sqrt{49 + 9} > 7$

Since, $\sqrt{52} < \sqrt{58}$, then the answer is B.

28) **Answer: C**

To find the area of the shaded region subtract smaller circle from bigger circle.

S bigger – S smaller = π (r bigger)2 – π (r smaller)2

⇒ S bigger – S smaller = π (5)2 – π (3)2 ⇒ 25π – 9π = 16π

29) **Answer: D**

$\sqrt{x} = -5 \to x = 25$

then; $\sqrt{x} - 4 = \sqrt{25} - 4 = 5 - 4 = 1$ and $\sqrt{3x + 6} = \sqrt{75 + 6} = \sqrt{81} = 9$ Then:

$(\sqrt{3x + 6}) + (\sqrt{x} - 4) = 9 + 1 = 10$

30) **Answer: 98.**

$a = 14 \Rightarrow$ area of the triangle is

$= \frac{1}{2}(14 \times 14) = \frac{196}{2} = 98 \ cm^2$

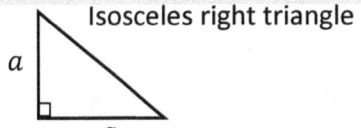

31) **Answer: A**

x is directly proportional to the square of y. Then: $x = cy^2$

$63 = c(3)^2 \to 63 = 9c \to c = \frac{63}{9} = 7$

The relationship between x and y is: $x = 7y^2$, $x = 175$

$175 = 7y^2 \to y^2 = \frac{175}{7} = 25 \to y = 5$

32) **Answer: C**

$\alpha = 180° - 122° = 58°$

$\beta = 180° - 118° = 62°$

$x + \alpha + \beta = 180° \to x = 180° - 58° - 62° = 60°$

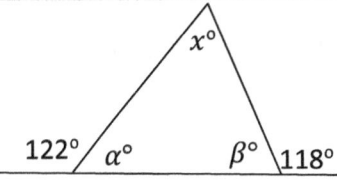

33) Answer: 3

Use formula of rectangle prism volume.

V = (length) (width) (height) ⇒ 1,344 = (32) (14) (height) ⇒ height = 1,344 ÷ 448 = 3

34) Answer: C

The amount of money that jack earns for one hour: $\frac{\$500}{25} = \$20$

Number of additional hours that he works to make enough money is: $\frac{\$920-\$500}{1.5 \times \$20} = 14$

Number of total hours is: $20 + 14 = 34$

35) Answer: C

Let's find the mean (average), mode and median of the number of cities for each type of pollution.

Number of cities for each type of pollution: 6, 3, 2, 9, 4

average (mean) = $\frac{sum\ of\ terms}{number\ of\ terms} = \frac{6+3+2+9+4}{5} = \frac{24}{5} = 4.8$

Median is the number in the middle. To find median, first list numbers in order from smallest to largest: 2, 3, 4, 6, 9

Median of the data is 4.

Mode is the number which appears most often in a set of numbers. Therefore, there is no mode in the set of numbers.

Median < Mean, then, $c < a$

36) Answer: B

Let the number of cities should be added to type of pollutions B be x. Then:

$\frac{x+3}{9} = 1 \rightarrow x + 3 = 9 \times 1 \rightarrow x + 3 = 9 \rightarrow x = 6$

37) Answer: A

Percent of cities in the type of pollution A: $\frac{6}{10} \times 100 = 60\%$

Percent of cities in the type of pollution C: $\frac{2}{10} \times 100 = 20\%$

Percent of cities in the type of pollution E: $\frac{4}{10} \times 100 = 40\%$

38) Answer: C

$AB = 6$, And $AC = 8$

$BC = \sqrt{6^2 + 8^2} = \sqrt{36 + 64} = \sqrt{100} = 10$

Perimeter $= 6 + 8 + 10 = 24$; Area $= \frac{6 \times 8}{2} = 24$

In this case, the ratio of the perimeter of the triangle to its area is: $\frac{24}{24} = 1$

If the sides AB and AC become half longer, then: $AB = 3$, And $AC = 4$

$BC = \sqrt{3^2 + 4^2} = \sqrt{9 + 16} = \sqrt{25} = 5$

Perimeter $= 3 + 4 + 5 = 12$; Area $= \frac{3 \times 4}{2} = 3 \times 2 = 6$

In this case the ratio of the perimeter of the triangle to its area is: $\frac{12}{6} = 2$

39) Answer: C

The capacity of a red box is 30% bigger than the capacity of a blue box and it can hold 78 books. Therefore, we want to find a number that 30% bigger than that number is 78. Let x be that number. Then:

$1.30 \times x = 78$, Divide both sides of the equation by 1.3. Then: $x = \frac{78}{1.30} = 60$

40) Answer: D

$\$15 \times 6 = \$90$

Petrol use: $3 \div 2 = 1.5$. $6 \times 1.5 = 9$ liters

Petrol cost: $9 \times \$2.2 = \$19.8$

Money earned: $\$90 - \$19.8 = \$70.2$

"End"

www.ingramcontent.com/pod-product-compliance
Lightning Source LLC
Chambersburg PA
CBHW081111080526
44587CB00021B/3547